T0214000

SpringerBriefs in Applied Sciences and Technology

SpringerBriefs present concise summaries of cutting-edge research and practical applications across a wide spectrum of fields. Featuring compact volumes of 50 to 125 pages, the series covers a range of content from professional to academic.

Typical publications can be:

- A timely report of state-of-the art methods
- An introduction to or a manual for the application of mathematical or computer techniques
- A bridge between new research results, as published in journal articles
- A snapshot of a hot or emerging topic
- An in-depth case study
- A presentation of core concepts that students must understand in order to make independent contributions

SpringerBriefs are characterized by fast, global electronic dissemination, standard publishing contracts, standardized manuscript preparation and formatting guidelines, and expedited production schedules.

On the one hand, **SpringerBriefs in Applied Sciences and Technology** are devoted to the publication of fundamentals and applications within the different classical engineering disciplines as well as in interdisciplinary fields that recently emerged between these areas. On the other hand, as the boundary separating fundamental research and applied technology is more and more dissolving, this series is particularly open to trans-disciplinary topics between fundamental science and cngineering.

Indexed by EI-Compendex, SCOPUS and Springerlink.

More information about this series at http://www.springer.com/series/8884

Afonso Botelho · Baltazar Parreira ·
Paulo N. Rosa · João Miranda Lemos

Predictive Control
for Spacecraft Rendezvous

 Springer

Afonso Botelho
Elecnor/DEIMOS
Madrid, Spain

Baltazar Parreira
Elecnor/DEIMOS
Lisbon, Portugal

Paulo N. Rosa
Elecnor/DEIMOS
Lisbon, Portugal

João Miranda Lemos ⓘ
Control of Dynamical Systems
INESC-ID
Lisbon, Portugal

ISSN 2191-530X   ISSN 2191-5318 (electronic)
SpringerBriefs in Applied Sciences and Technology
ISBN 978-3-030-75695-6   ISBN 978-3-030-75696-3 (eBook)
https://doi.org/10.1007/978-3-030-75696-3

© The Author(s), under exclusive license to Springer Nature Switzerland AG 2021
This work is subject to copyright. All rights are solely and exclusively licensed by the Publisher, whether the whole or part of the material is concerned, specifically the rights of translation, reprinting, reuse of illustrations, recitation, broadcasting, reproduction on microfilms or in any other physical way, and transmission or information storage and retrieval, electronic adaptation, computer software, or by similar or dissimilar methodology now known or hereafter developed.
The use of general descriptive names, registered names, trademarks, service marks, etc. in this publication does not imply, even in the absence of a specific statement, that such names are exempt from the relevant protective laws and regulations and therefore free for general use.
The publisher, the authors and the editors are safe to assume that the advice and information in this book are believed to be true and accurate at the date of publication. Neither the publisher nor the authors or the editors give a warranty, expressed or implied, with respect to the material contained herein or for any errors or omissions that may have been made. The publisher remains neutral with regard to jurisdictional claims in published maps and institutional affiliations.

This Springer imprint is published by the registered company Springer Nature Switzerland AG
The registered company address is: Gewerbestrasse 11, 6330 Cham, Switzerland

Preface

This book addresses the design of a model predictive control algorithm for performing spacecraft rendezvous manoeuvres. Although hundreds of rendezvous missions have been successfully carried out, the development of new guidance and control algorithms remains an active area of research, motivated by the demand for improved efficiency, safety, and autonomy of these manoeuvres, which this book attempts to consolidate. The book is accessible to those new to the topics covered, regarding both orbital rendezvous and model predictive control but also presents compelling subjects for researchers and professionals in the aerospace industry, including some contributions to this area of research. In addition, the book is a useful complement to courses on model-based predictive control.

The present work was initially developed and adapted from the first author's (A. Botelho) MSc dissertation, performed under the supervision of the remaining authors. The authoring team counts with the experience of a full professor (J. M. Lemos) at the University of Lisbon with expert knowledge in optimal control, as well as two professionals (B. Parreira and P.N. Rosa) of the aerospace company Elecnor Deimos, a European Space Agency partner and a strong player in the European aerospace industry. The thesis was motivated as a feasibility study on the application of the aforementioned techniques to the ESA PROBA-3 mission Rendezvous Experiment (RVX), led by Deimos Engenharia.

The authors acknowledge the European Space Agency (ESA) for granting permission for use of the CLGADR high-fidelity simulator, developed by Deimos under ESA contract No. 4000111160/14/NL/MH, for validation of results. The views expressed herein can in no way be taken to reflect the official opinion of the European Space Agency.

Part of this work was supported by project HARMONY, Distributed Optimal Control for Cyber-Physical Systems Applications, financed by FCT (Portugal) under contract AAC nº2/SAICT/2017-031411.

Lisbon, Portugal
2020

Afonso Botelho
Baltazar Parreira
Paulo N. Rosa
João Miranda Lemos

Contents

Chapter 1
Introduction

Orbital rendezvous is a procedure in which two separate spacecraft meet at the same orbit, as illustrated in Fig. 1.1, thereby approximately matching their orbital velocity and position [1]. Such manoeuvres allowed for the feasibility of the Apollo moon landing missions, with the rendezvous of the Lunar Excursion Module with the Command Module in lunar orbit, and for the construction and resupply of modular space stations, such as Mir and the International Space Station. Other applications include, for example, the exploration of smaller celestial bodies, such as asteroids, comets and small moons, the in-orbit servicing of satellites, for instance, the multiple repair missions to the Hubble Space Telescope, or the active removal of space debris. Often, the rendezvous process is followed by a docking or berthing procedure, that results in the physical connection of the two spacecraft.

The first attempt at rendezvous was performed in the Gemini 4 manned mission in 1965, which was unsuccessful due to the method of approach being simply 'point-and-shoot', resulting in a further separation of the spacecraft. This revealed the challenge in performing space rendezvous, and proved that the relative orbital dynamics involving the two spacecraft must be taken into consideration. Since then, rendezvous missions have been performed successfully hundreds of times, both by manned and unmanned spacecraft, and using various different guidance and control methods. In this context, this book addresses the use of Model Predictive Control (MPC) [2] for performing rendezvous manoeuvres, which is a widely successful optimal control strategy that naturally considers the system dynamics and can handle various operational constraints. The use of MPC for this purpose can grant more autonomy to the spacecraft and improve the optimality of the approach trajectories, when compared to the traditional techniques.

The literature for MPC applied to rendezvous is now quite considerable, and this topic remains an active area of research. Despite these facts, MPC has been tested in real spaceflight only once, to the best of the authors' knowledge, by the

© The Author(s), under exclusive license to Springer Nature Switzerland AG 2021
A. Botelho et al., *Predictive Control for Spacecraft Rendezvous*,
SpringerBriefs in Applied Sciences and Technology,
https://doi.org/10.1007/978-3-030-75696-3_1

Fig. 1.1 Illustration of the orbital rendezvous manoeuvre

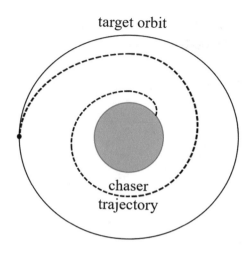

PRISMA mission [3]. Although this test was made in a formation flying context, not rendezvous, the underlying principles are identical. The main difficulty with the use of MPC for a real rendezvous mission is that it requires a considerable online computational effort, that can prove to be a challenge given the typically limited computing power available on board. Furthermore, there is not yet a standard strategy for granting robustness in the presence of disturbances possibly interfering with a rendezvous mission that is both feasible to implement in real time and maintains good operational performance, and thus more research into this topic is required.

1.1 Problem Formulation

A spacecraft rendezvous mission generally adheres to the following sequence of events: launch, phasing, far-range rendezvous, close-range rendezvous and mating [1]. The launch phase ends with orbital insertion, nearly coplanar with the target orbit and typically at a lower altitude and behind the target, and is completely out of the scope of this work. Phasing consists of small corrections to the orbit parameters, and of passive waiting that takes advantage of the different orbital periods, in order to allow the launched spacecraft to catch up the target. This phase can last a few days and does not require great precision, with correction manoeuvres being performed in open loop without the need for the use of MPC. The rendezvous process starts then with the far-range phase when relative navigation is possible which typically within a range of a few tens of kilometres with respect to the target. This phase ends and close-range rendezvous begins when the relative distance requires safety-critical manoeuvres, typically at a few kilometres. Thus, these two phases are the ones that can benefit from the use of MPC to perform the approach manoeuvres, with the latter phase being the focus of this book. The mating phase (docking or berthing) typically has a very different set of requirements and constraints and thus is also out of scope for this book.

An on-board automatic control system for a spacecraft contains three sub-systems tasked with the execution of thrust manoeuvres: guidance, navigation and control (GNC) [1]. The guidance system generates the reference trajectory and spacecraft attitude; navigation provides state measurements and estimates; control commands the force and torque necessary to drive the spacecraft to the desired state. MPC can simultaneously handle both guidance and control functions and navigation is not considered in this book. Furthermore, because the translational and attitude control are typically decoupled in the far and close-range rendezvous phases [1], only translational control is addressed in this book. Finally, it is remarked that, while all real rendezvous and formation flying missions have been performed in circular or near-circular target orbits, there is the motive to consider elliptical target orbits, which imply an increased difficulty due to the dynamics becoming time-varying and, therefore, more complex. For example, for the future European Space Agency PROBA-3 mission, the spacecraft will be placed in a highly elliptical orbit in order to validate formation flying and rendezvous technology.

1.2 State of the Art

This book covers several different research areas, and thus we will address the essentials of the state of the art for these separately. MPC was first introduced in the 1960s and is now a very mature framework, with an extensive theoretical basis [2] and a vast history of successful applications, mostly in the process industry [4]. It remains an active area of research, with recent work being dedicated to the application of MPC to specific problems, including the rendezvous scenario [5]. Research has also been devoted to improving the real-time feasibility of MPC, with the design of new optimization algorithms that exploit the MPC problem structure, for example, [6–8] and with the further development of the popular Explicit MPC framework [9, 10]. New sub-fields of MPC have also emerged in recent years, such as Distributed MPC [11, 12], Hybrid MPC [13], Adpative MPC [14], Stochastic MPC [15] or Neural Network MPC [16], among several others.

Concerning relative orbital mechanics, although the nonlinear dynamics can be easily derived from Newton's laws [17], these differential equations do not have a closed-form solution, which limits their usefulness to an on-board and real-time environment. Thus, research into this topic is still active today, dedicated to determining approximated dynamics with a closed-form solution, necessary for any rendezvous strategy. A set of linearized equations for the relative motion represented in Cartesian coordinates in a local non-inertial frame of reference and for a circular orbit was first derived by Hill in 1878 [18]. They were first applied and solved in the context of orbital rendezvous in 1959, most famously by Clohessy and Wiltshire [19], although this solution is only accurate for near-circular orbits. The equations were extended to elliptic orbits by De Vries in 1963 [20], and simplified via a change of the independent variable to the true anomaly. They were then solved and applied to spacecraft rendezvous in elliptic orbits by Tschauner and Hempel in 1965 [21], after whom the simplified equations became known. In 1998, Carter presented a simpler solution to the Tschauner–Hempel equations in the form of a state transition matrix [22], which is valid for any orbit eccentricity. Later, in 2002, Yamanaka and Ankersen introduced a computationally simpler state transition matrix [23], although this description is only valid for circular or elliptical orbits. Ankersen later complemented this solution by including a forced regime with a constant force discretization [24], which can be used for spacecraft control.

More recent work focuses on formulating linearized models that include different perturbations. The Hill–Clohessy–Wiltshire equations were extended to include the J_2 perturbation in [25], and atmospheric drag in [26], although these remain valid only for near-circular target orbits. A dynamic model which includes J_2 and is valid for elliptical target orbits was later presented in [27], and another which also includes atmospheric drag was presented in [28].

While all the previously mentioned dynamic models are based on Cartesian coordinates, another approach commonly used for spacecraft rendezvous and formation control is based on the Keplerian orbital elements. Lagrange's Planetary Equations and the Gauss Variational Equations [29] model the effect of, respectively, conservative and arbitrary local-frame perturbations, e.g. gravitational perturbations or a controlled thrust, on the orbital elements of a satellite. From these, linearized models of the relative dynamics of two satellites can be formulated in terms of Relative Orbital Elements (ROEs), rather than the relative position and velocity typically utilized. The ROEs may be arbitrarily defined from the target and chaser absolute orbital elements, where the most straightforward formulation is a simple difference of the elements of the two satellites [30–33]. The resulting linear equations present some advantages with respect to Cartesian-based models, such as remaining accurate for larger relative distances since they are linearized around the target orbital elements rather than the target position, readily include the target eccentricity and more easily allow the inclusion of higher order potential models and other perturbations. More recent dynamic models based on different ROE formulations have been presented by D'Amico et al. [34–36], which more easily allow for a stable and passively safe trajectory design, and include the J_2 perturbation and atmospheric drag [37].

The Yamanaka–Ankersen state transition matrix is considered to be the state-of-the-art solution for use in the design of rendezvous missions in elliptic orbits, due to its matureness, inclusion of the target eccentricity, relative simplicity when compared to other models and its representation in Cartesian coordinates, which are more intuitive than other state representations. Nevertheless, research to determine the different and improved linearized relative dynamic models continues. References [38, 39] present exhaustive surveys on the models currently available in the literature, and perform accuracy comparisons between the models, as well as runtime comparisons in the case of the latter.

Current state-of-the-art methods for rendezvous guidance and control rely on commanding the spacecraft to follow a sequence of waypoints, typically defined offline during mission analysis, using simple manoeuvres with analytical solutions, e.g.

two boosts or straight-line approaches, possibly with some limited mid-correction manoeuvring [1, 40, 41]. Such methods have a low computational burden compatible with the real-time environment they are required in, are robust and have great maturity and history of successful applications, at the sacrifice of optimality and autonomy, however.

The application of optimal control theory to rendezvous problems was pioneered in the 1950s by Lawden [42], in what culminated in the *primer vector* theory. Although technically the underlying theory is the calculus of variations and not optimal control, Lawden formulated first-order conditions for optimal spacecraft trajectories. Since then, several books on spacecraft optimal control have been published, consolidating the field, e.g. [43–45].

Despite the current relative *status quo*, research into the rendezvous guidance and control problem using different methods is vast, namely using optimal control methods. However, these works are based on indirect numerical methods (various forms of Pontryagin's principle), which present several practical limitations [46]. Research on the application of direct methods, such as MPC or pseudospectral methods [47], has only begun in more recent decades, driven by the increasing available computational capability, and motivated by an increasing necessity for optimality and autonomy.

One of the first applications of MPC to the rendezvous problem was by Richards and How [48], where the basic formulations employed by most of the literature that followed were presented, namely Fixed-Horizon (FH) MPC and Variable-Horizon (VH) MPC, although the optimization formulations had been introduced earlier [49] but not in an rendezvous MPC context. These formulations explicitly minimize the fuel for the rendezvous manoeuvres and, in the case of the former, express the optimization problem as a linear program, that can allow for a feasible online computation time. Thus, current research is mostly dedicated to extending these formulations, for example, to grant robustness, in the presence of the many disturbances and perturbations interfering in a rendezvous mission, while ensuring convergence, constraint satisfaction and performance. A more in-depth state-of-the-art review for this topic is available along with Chap. 4 of this book, with special attention to robustness techniques.

1.3 Contributions

The book starts by featuring a basic introduction to general MPC theory, with simple toy problems and experiments that demonstrate the capabilities of this control method, and that can serve as a practical tutorial for those uninitiated on this topic. It also contains an introduction to relative orbital mechanics, with a derivation of the approximated dynamic model, and featuring several simulations with the explanations required to fully understand the rendezvous dynamics and manoeuvres.

Our contributions to MPC applied to rendezvous include considering rendezvous manoeuvres in highly elliptical orbits, which is uncommon in the literature, especially with an eccentricity as high as that considered here. With this condition in mind, a new approach for sampling the dynamics for the prediction horizon is proposed here, based on constant eccentric anomaly sampling intervals, which deals better with the fact that the dynamics are highly time-varying for such a highly elliptical orbit. We compare the Finite-Horizon and VH MPC formulations, regarding performance and computational complexity, and compare them with the traditional two-impulse transfer approach used in traditional rendezvous guidance algorithms.

A new method for formulating obstacle avoidance constraints and passive safety constraints is also presented, that relies on iterative linear optimization and which allows for the feasible inclusion of this constraint in a real-time application while maintaining optimality. Finally, the book contributes to new robustness techniques, with the use of a terminal quadratic controller for a more accurate and robust final braking manoeuvre, and the dynamic relaxation of the terminal constraint, in order to keep the control sparse and avoid the overcorrection of disturbances and waste of fuel. The latter techniques are demonstrated in an industrial high-fidelity simulation environment, considering the conditions of the ESA PROBA-3 rendezvous experiment (RVX).

1.4 Book Outline

In Chap. 2, the book covers general MPC theory, with a focus on MPC for linear system models, given the context of this book. The basic general formulation is discussed, and some specific techniques are also presented, such as reference tracking and the use of different cost functions. This chapter also features several simulations that show the effect of different cost functions and the choice of controller parameters, with some consideration also given to the computational performance.

The relative orbital dynamics between two satellites, that is crucial to understanding the design of a rendezvous mission, is presented in Chap. 3. The book introduces and derives linearized models of the relative dynamics, that will then be used for MPC in the next chapter. Several simulations of the relative motion between two satellites are also presented, both in the circular and elliptic orbit cases and the non-intuitive free-drift motions and thrust manoeuvres are explained.

Finally, in Chap. 4 the MPC framework is applied to the rendezvous problem. The book starts by considering the most naive approach and develops stepwise toward the ideal formulation. It then considers the presence of disturbances and provides a short literature review on robust techniques in MPC for rendezvous.

References

1. W. Fehse, *Automated Rendezvous and Docking of Spacecraft* (Cambridge University Press, 2003). ISBN: 0521824923

2. J. Rawlings, D. Mayne, M. Diehl, *Model Predictive Control: Theory, Computation, and Design*, 2nd edn. (Nob Hill Publishing, 2017)

3. P. Bodin, R. Noteborn, R. Larsson, C. Chasset, System test results from the GNC experiments on the PRISMA in-orbit test bed. Acta Astronaut. **68**, 862–872 (2011). ISSN: 0094-5765

4. J. Han, Y. Hu, S. Dian, The state-of-the-art of model predictive control in recent years. IOP Conf. Ser.: Mater. Sci. Eng. **428**, 012035 (2018)

5. E.N. Hartley, A tutorial on model predictive control for spacecraft rendezvous in *2015 European Control Conference (ECC)*, July 2015 (2015), pp. 1355–1361

6. C.V. Rao, S.J. Wright, J.B. Rawlings, Application of interior-point methods to model predictive control. J. Optim. Theory Appl. **99**, 723–757 (1998)

7. A. Domahidi, A.U. Zgraggen, M.N. Zeilinger, M. Morari, C.N. Jones, Efficient interior point methods for multistage problems arising in receding horizon control, in *2012 IEEE 51st IEEE Conference on Decision and Control (CDC)* (2012), pp. 668–674

8. D. Liao-McPherson, M. Huang, I. Kolmanovsky, A regularized and smoothed Fischer-Burmeister method for quadratic programming with applications to model predictive control. IEEE Trans. Autom. Control. (2018)

9. A. Bemporad, M. Morari, V. Dua, E.N. Pistikopoulos, The explicit linear quadratic regulator for constrained systems. Automatica **38**, 3–20 (2002)

10. A. Alessio, A. Bemporad, *Nonlinear Model Predictive Control* (Springer, Berlin, 2009), pp. 345–369

11. E. Camponogara, D. Jia, B.H. Krogh, S. Talukdar, Distributed model predictive control. IEEE Control Syst. Mag. **22**, 44–52 (2002)

12. A.N. Venkat, I.A. Hiskens, J.B. Rawlings, S.J. Wright, Distributed MPC strategies with application to power system automatic generation control. IEEE Trans. Control Syst. Technol. **16**, 1192–1206 (2008)

13. E. Camacho, D. Ramirez, D. Limon, D. Muñoz de la Peña, T. Alamo, Model predictive control techniques for hybrid systems. Annu. Rev. Control **34**, 21–31 (2010), http://www.sciencedirect.com/science/article/pii/S1367578810000040. ISSN: 1367-5788

14. V. Adetola, D. DeHaan, M. Guay, Adaptive model predictive control for constrained nonlinear systems. Syst. Control Lett. **58**, 320–326 (2009), http://www.sciencedirect.com/science/article/pii/S0167691108002120. ISSN: 0167-6911

15. A. Mesbah, Stochastic model predictive control: an overview and perspectives for future research. IEEE Control Syst. Mag. **36**, 30–44 (2016)

16. A. Draeger, S. Engell, H. Ranke, Model predictive control using neural networks. IEEE Control Syst. Mag. **15**, 61–66 (1995)

17. W.E. Wiesel, *Spaceflight Dynamics*, 3rd edn. (Aphelion Press, 2010)

18. G.W. Hill, Researches in the lunar theory. Am. J. Math. **1**, 5-26. (1878). ISSN: 00029327, 10806377

19. W.H. Clohessy, R.S. Wiltshire, Terminal guidance system for satellite rendezvous. J. Aerosp. Sci. **27**, 653–658 (1960)

20. J.P. De Vries, Elliptic elements in terms of small increments of position and velocity components. AIAA J. **1**, 2626–2629 (1963)

21. J. Tschauner, P. Hempel, Rendezvous with a target in an elliptical orbit. Astronaut. Acta **11**, 104–109 (1965)

22. T.E. Carter, State transition matrices for terminal rendezvous studies: brief survey and new example. J. Guid. Control Dyn. **21**, 148–155 (1998)

23. K. Yamanaka, F. Ankersen, New state transition matrix for relative motion on an arbitrary elliptical orbit. J. Guid. Control Dyn. **25**, 60–66 (2002)

24. F. Ankersen, Guidance, navigation, control and relative dynamics for spacecraft proximity maneuvers. Ph.D. thesis, Institut for Elektroniske Systemer (2010). ISBN: 9788792328724

25. S.A. Schweighart, R.J. Sedwick, High-fidelity linearized J model for satellite formation flight. J. Guid. Control Dyn. **25**, 1073–1080 (2002)

26. R. Bevilacqua, M. Romano, Rendezvous maneuvers of multiple spacecraft using differential drag under J2 perturbation. J. Guid. Control Dyn. **31**, 1595–1607 (2008)

27. C. Wei, S.-Y. Park, C. Park, Linearized dynamics model for relative motion under a J2-perturbed elliptical reference orbit. Int. J. Non-Linear Mech. **55**, 55–69 (2013)
28. L. Cao, H. Li, Linearized J2 and atmospheric drag model for control of inner-formation satellite system in elliptical orbits. J. Dyn. Syst. Meas. Control **138** (2016)
29. K. Alfriend, S.R. Vadali, P. Gurfil, J. How, L. Breger, *Spacecraft Formation Flying: Dynamics, Control and Navigation* (Elsevier, 2009)
30. H. Schaub, S.R. Vadali, J.L. Junkins, K.T. Alfriend, Spacecraft formation flying control using mean orbit elements. J. Astronaut. Sci. **48**, 69–87 (2000)
31. K. Alfriend, Nonlinear considerations in satellite formation flying, in *AIAA/AAS Astrodynamics Specialist Conference and Exhibit* (2002), p. 4741
32. D.-W. Gim, K.T. Alfriend, Satellite relative motion using differential equinoctial elements. Celest. Mech. Dyn. Astron. **92**, 295–336 (2005)
33. L. Breger, J.P. How, Gauss's variational equation-based dynamics and control for formation flying spacecraft. J. Guid. Control Dyn. **30**, 437–448 (2007)
34. S. D'Amico, Relative orbital elements as integration constants of Hill's equations. DLR, TN, 05-08 (2005)
35. O. Montenbruck, M. Kirschner, S. D'Amico, S. Bettadpur, E/I-vector separation for safe switching of the GRACE formation. Aerosp. Sci. Technol. **10**, 628–635 (2006)
36. S. D'Amico, Autonomous formation flying in low earth orbit. Ph.D. thesis, TU Delft (2010)
37. A.W. Koenig, T. Guffanti, S. D'Amico, New state transition matrices for spacecraft relative motion in perturbed orbits. J. Guid. Control Dyn. **40**, 1749–1768 (2017)
38. K. Alfriend, H. Yan, Evaluation and comparison of relative motion theories. J. Guid. Control Dyn. **28**, 254–261 (2005)
39. J. Sullivan, S. Grimberg, S. D'Amico, Comprehensive survey and assessment of spacecraft relative motion dynamics models. J. Guid. Control Dyn. **40**, 1837–1859 (2017)
40. F. Liu, S. Lu, Y. Sun, *Guidance and Control Technology of Spacecraft on Elliptical Orbits* (Springer, Berlin, 2019).
41. Y. Xie, C. Chen, T. Liu, M. Wang, *Guidance, Navigation, and Control for Spacecraft Rendezvous and Docking: Theory and Methods* (Springer, Berlin, 2021).
42. D.F. Lawden, Fundamentals of space navigation. J. Br. Interplanet. Soc. **13**, 87–101 (1954)
43. T.N. Edelbaum, Optimal space trajectories. Technical report, ANALYTICAL MECHANICS ASSOCIATES INC JERICHO NY (1969)
44. J.-P. Marec, *Optimal Space Trajectories* (Elsevier, 1979). ISBN: 0-444-41812-1
45. J.M. Longuski, J.J. Guzmán, J.E. Prussing, *Optimal Control with Aerospace Applications* (Springer, Berlin, 2014).
46. E. Trélat, *Contrôle Optimal: Thèorie & Applications* (Vuibert, Paris, 2005).
47. I.M. Ross, M. Karpenko, A review of pseudospectral optimal control: from theory to flight. Annu. Rev. Control **36**, 182–197 (2012)
48. A. Richards, J. How, Performance evaluation of rendezvous using model predictive control, in *AIAA Guidance, Navigation, and Control Conference and Exhibit*, November 2003 (2003)
49. A. Bemporad, M. Morari, Control of systems integrating logic, dynamics, and constraints. Automatica **35**, 407–427 (1999)

Chapter 2
Model Predictive Control

Model Predictive Control (MPC) is a control design method based on iterative online optimization [1]. The strategy is to obtain a control decision by solving an optimization problem which factors in future states of the system in a finite horizon, predicted using a (generally discrete) system model. Figure 2.1 illustrates this approach.

At each time step, the problem is solved with the most recent state measurement or estimate as to the initial condition for the prediction, and a control strategy for future steps within the prediction horizon is obtained. The first control value in the obtained sequence is applied to the plant, and the problem is solved again in the next time step, with an updated state and with the prediction horizon shifted forward. For this reason, this method is also known as Moving/Receding-Horizon Control.

Since MPC is formulated as an optimization problem, it allows for the inclusion of control and state constraints. The possibility to explicitly include constraints is a powerful tool and one of the major advantages of MPC in respect to other control methods since it allows to limit the control action and to model complex state restrictions, such as safety constraints. Furthermore, MPC naturally considers the system dynamics and can handle multivariate systems. It may also feature the use of nonlinear system models, that can generate better state predictions.

By definition, the MPC strategy requires that an optimization problem is solved online, at each time step. The computation time of the MPC problem depends on many factors, such as the order of the system model, linearity, the complexity of the control and state constraints and the length of the prediction horizon. The optimal control action must be computed and applied to the plant before the next sample, and thus the problem is required to be solved faster than the system sampling time, which makes its implementation infeasible in fast systems. This computational requirement is the major limitation for MPC, although modern technology and methods allow for MPC to be implemented in increasingly more complex systems, such as those in the aerospace industry.

Other major issues which are still the subject of current research are stability [2] and robustness [1]. Many stability proofs rely on imposing a terminal constraint and

© The Author(s), under exclusive license to Springer Nature Switzerland AG 2021
A. Botelho et al., *Predictive Control for Spacecraft Rendezvous*,
SpringerBriefs in Applied Sciences and Technology,
https://doi.org/10.1007/978-3-030-75696-3_2

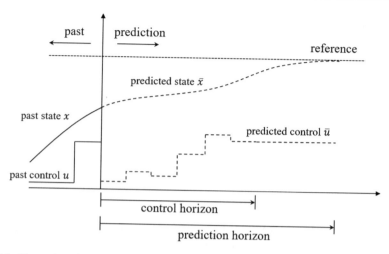

Fig. 2.1 Illustration of the Model Predictive Control strategy

obtaining an equivalence to an infinite horizon, or modified infinite horizon problem and using a Lyapunov-type argument. See Sect. 2.1 of [1] for an overview.

In this chapter, we first introduce the basic concepts of MPC theory in Sect. 2.1, and then discuss different methods, including MPC for linear systems in Sect. 2.2 and briefly for nonlinear systems in Sect. 2.3. Section 2.4 presents a commonly used sub-optimal complexity reduction technique. In Sect. 2.5, we show several MPC experiments with a linear two-dimensional pure inertial system, modelling an inertial vehicle in free-space.

2.1 Model Predictive Control Formulation

MPC theory is traditionally formulated in discrete time, and so is it here. A discrete-time system, with state variables x, inputs u and outputs y, is generally described by the difference equation

$$x^+ = f(x, u), \tag{2.1}$$

where x^+ is the system state at the next time step and $f(x, u)$ is the system model, and by the output equation

$$y = g(x, u), \tag{2.2}$$

where $g(x, u)$ is the sensor model. If models f and g are linear, the system can instead be described by the linear state model

$$x^+ = Ax + Bu, \tag{2.3}$$
$$y = Cx + Du, \tag{2.4}$$

where usually $D = 0$, except for systems with an instantaneous response. For literature on discrete-time systems and digital control, see [3].

Model Predictive Control solves an open-loop optimal control problem at every time-step. At a time instant t the following optimization problem is solved:

$$\min_{\substack{\bar{u}_t, \ldots, \bar{u}_{t+N-1} \\ \bar{x}_t, \ldots, \bar{x}_{t+N}}} \quad \sum_{i=0}^{N-1} l(\bar{x}_{t+i}, \bar{u}_{t+i}) + V_f(\bar{x}_{k+N}), \tag{2.5a}$$

$$\text{s.t.} \quad \bar{x}_t = x_t, \tag{2.5b}$$
$$\bar{x}^+ = f(\bar{x}, \bar{u}), \tag{2.5c}$$
$$\bar{x}_k \in \mathcal{X}_k, \ k = t, \ldots, t + N, \tag{2.5d}$$
$$\bar{u}_k \in \mathcal{U}_k, \ k = t, \ldots, t + N - 1, \tag{2.5e}$$

where $\bar{x}$ and $\bar{u}$ are the predictions of x and u, N is the length of the prediction horizon and $l(\cdot, \cdot)$ and $V_f(\cdot)$ are instantaneous cost functions, often known as the Lagrangian/stage and Mayer/terminal terms, respectively. The minimization is subject to ('s.t.') constraints (2.5b–e). The constraint in (2.5b) sets the initial condition for the prediction, and the state and control predictions are subject to the system model in constraint (2.5c). The sets $\mathcal{X}$ and $\mathcal{U}$ in (2.5d) and (2.5e) represent constraints on the state and control variables, respectively. Solving this open-loop problem yields an optimal control sequence $\bar{u}^*$, of which only the first is applied, meaning that $u_t = \bar{u}_t^*$. The problem is then solved again at the next time-step, at $t + 1$, with the updated state measurement/estimate x_{t+1} that is the response to the applied control, thus closing the control loop.

For literature on more MPC theory, such as stability of the closed-loop system, see, for example, the book by Rawlings et al. [1].

2.2 Linear Model Predictive Control

Linear MPC refers to the control of linear systems. For a linear system, the MPC formulation is

$$\min_{\substack{\bar{u}_0,\dots,\bar{u}_{N-1} \\ \bar{x}_0,\dots,\bar{x}_N}} \sum_{i=0}^{N-1} l(\bar{x}_i, \bar{u}_i) + V_f(\bar{x}_N), \tag{2.6a}$$

$$\text{s.t.} \quad \bar{x}_0 = x_0, \tag{2.6b}$$

$$\bar{x}_{k+1} = A\bar{x}_k + B\bar{u}_k, \ k = 0, \dots, N-1, \tag{2.6c}$$

$$\bar{x}_k \in \mathcal{X}_k, \ k = 0, \dots, N, \tag{2.6d}$$

$$\bar{u}_k \in \mathcal{U}_k, \ k = 0, \dots, N-1, \tag{2.6e}$$

where the constraint (2.6c) is now the linear state model of the system. Since this constraint is linear, the problem is easier and faster to solve, in comparison to one with a nonlinear system. For simplicity, t has been set to 0 in the above formulation.

A common and effective cost for the MPC optimization problem is the quadratic cost, where the cost functions become

$$l(x, u) = x^\top Q x + u^\top R u,$$
$$V_f(x) = x^\top Q_f x, \tag{2.7}$$

where Q and Q_f are positive semidefinite matrices and R is a positive definite matrix. These cost matrices are used to tune the controller: increasing the elements in R relative to Q and Q_f increases the penalization of the control variable in the cost function, and so the optimal solution will have limited actuator action. This formulation, called 'regulator', drives the state to the origin. A common choice for the terminal state matrix Q_f is the solution to the algebraic Riccati equation since in some cases it guarantees closed-loop stability [2].

For an easy and efficient implementation of linear quadratic MPC, it is convenient to adopt a matrix representation. Concatenating the predicted state and control variables into X and U, we have the following matrix equation that satisfies the predictive model:

$$\underbrace{\begin{bmatrix} \bar{x}_0 \\ \bar{x}_1 \\ \vdots \\ \bar{x}_N \end{bmatrix}}_{X} = \underbrace{\begin{bmatrix} 0 & \dots & \dots & 0 \\ A & \dots & 0 & \vdots \\ \vdots & \ddots & \vdots & \vdots \\ 0 & \dots & A & 0 \end{bmatrix}}_{\tilde{A}} \underbrace{\begin{bmatrix} \bar{x}_0 \\ \bar{x}_1 \\ \vdots \\ \bar{x}_N \end{bmatrix}}_{X} + \underbrace{\begin{bmatrix} 0 & \dots & 0 \\ B & \dots & 0 \\ \vdots & \ddots & \vdots \\ 0 & \dots & B \end{bmatrix}}_{\tilde{B}} \underbrace{\begin{bmatrix} \bar{u}_0 \\ \bar{u}_1 \\ \vdots \\ \bar{u}_{N-1} \end{bmatrix}}_{U} + \underbrace{\begin{bmatrix} I \\ 0 \\ \vdots \\ 0 \end{bmatrix}}_{E} x_t. \tag{2.8}$$

Matrices $\tilde{A}$ and $\tilde{B}$ are augmented system model matrices and matrix E ensures the initial condition of the prediction. We also define the augmented cost matrices

$$\tilde{Q} = \begin{bmatrix} Q & \dots & 0 \\ \vdots & \ddots & \vdots \\ 0 & \dots & Q_f \end{bmatrix}, \quad \tilde{R} = \begin{bmatrix} R & \dots & 0 \\ \vdots & \ddots & \vdots \\ 0 & \dots & R \end{bmatrix}. \tag{2.9}$$

These augmented matrices can be easily generated by performing the Kronecker tensor product with the identity matrix.

With this matrix representation, the problem in (2.6) with the quadratic cost in (2.7) simplifies to

$$\min_{X,\, U} \quad X^\top \tilde{Q} X + U^\top \tilde{R} U, \tag{2.10a}$$

$$\text{s.t.} \quad X = \tilde{A} X + \tilde{B} U + E x(t), \tag{2.10b}$$

$$X \in \tilde{\mathcal{X}}, \tag{2.10c}$$

$$U \in \tilde{\mathcal{U}}, \tag{2.10d}$$

where $\tilde{\mathcal{X}}$ and $\tilde{\mathcal{U}}$ represent the state and control constraint sets over the whole prediction horizon.

It is sometimes useful to formulate the MPC problem with the system output, instead of the state. In this case, the MPC output form with quadratic cost is then

$$\min_{\substack{\bar{u}_0,\ldots,\bar{u}_{N-1} \\ \bar{y}_0,\ldots,\bar{y}_N}} \quad \sum_{i=0}^{N-1} \bar{y}_i^\top Q \bar{y}_i + \bar{u}_i^\top R \bar{u}_i + \bar{y}_N^\top Q_f \bar{y}_N, \tag{2.11a}$$

$$\text{s.t.} \quad \bar{x}_0 = x_t, \tag{2.11b}$$

$$\bar{x}_{k+1} = A\bar{x}_k + B\bar{u}_k, \ k = 0, \ldots, N-1, \tag{2.11c}$$

$$\bar{y}_k = C\bar{x}_k + D\bar{u}_k, \tag{2.11d}$$

$$\bar{y}_k \in \mathcal{Y}_k, \ k = 0, \ldots, N, \tag{2.11e}$$

$$\bar{x}_k \in \mathcal{X}_k, \ k = 0, \ldots, N, \tag{2.11f}$$

$$\bar{u}_k \in \mathcal{U}_k, \ k = 0, \ldots, N-1, \tag{2.11g}$$

where $\bar{y}$ is the predicted output and $\mathcal{Y}$ represents output constraints. This formulation now includes the system output equation in constraint (2.11d), as well as output constraints in (2.11e).

2.2.1 Reference Tracking

To drive the system to a reference set point x_{ref} instead of to the origin, the tracking error must penalized and so the instantaneous cost functions become

$$l(x, u) = (x - x_{ref})^\top Q(x - x_{ref}) + u^\top R u,$$
$$V_f(x) = (x - x_{ref})^\top Q_f(x - x_{ref}). \tag{2.12}$$

For systems without integral action, however, the optimal solution will not be at $x = x_{ref}$ for x_{ref} different than zero. For these systems, maintaining the state at a value different than the origin requires a constant non-zero control, which will weigh on the cost function and distance x from x_{ref}. Therefore, this formulation presents a static error for systems without integral action. To achieve an error-free reference tracking for these systems, a control reference u_{ref} must be added

$$l(x, u) = (x - x_{ref})^\top Q(x - x_{ref}) + (u - u_{ref})^\top R(u - u_{ref}). \tag{2.13}$$

The control reference u_{ref} must be the control value that in steady state makes the state be equal to the reference, and so from the system model we have, provided that the inverse of B exists,

$$u_{ref} = B^{-1}(I - A)x_{ref}. \tag{2.14}$$

It is remarked, however, that the state reference x_{ref} cannot be arbitrarily chosen since some states cannot be maintained in steady state. For example, a vehicle cannot maintain the same position while simultaneously having non-zero velocity. To check if the reference x_{ref} can be tracked in steady state, the result in (2.14) can be inserted back into the system model; if the state at the next time-step is not equal to the reference, then it is not admissible. For systems with integral action, u_{ref} will be zero for admissible state references.

The reference tracking formulation in matrix form is

$$\min_{X, U} \quad (X - X_{ref})^\top \tilde{Q}(X - X_{ref}) + (U - U_{ref})^\top \tilde{R}(U - U_{ref}), \tag{2.15a}$$

$$\text{s.t.} \quad X = \tilde{A}X + \tilde{B}U + Ex(t), \tag{2.15b}$$

$$U \in \tilde{\mathcal{U}}, \tag{2.15c}$$

$$X \in \tilde{\mathcal{X}}, \tag{2.15d}$$

where $X_{ref} = [x_{ref}^\top \ldots x_{ref}^\top]^\top$ and $U_{ref} = [u_{ref}^\top \ldots u_{ref}^\top]^\top$. Because in (2.14) the control reference is determined from the system model, this technique only completely eliminates the static error if the model is perfect.

An alternative way to achieve reference tracking without static error is by adding integral action to the controller. This can be performed by penalizing the control increment Δu between samples, instead of the full control action u. Thus, in steady state, the control action will remain constant and the increment Δu will be zero, eliminating the static error. The optimization problem then becomes

$$\min_{\substack{\bar{u}_0,\dots,\bar{u}_{N-1} \\ \Delta\bar{u}_0,\dots,\Delta\bar{u}_{N-1} \\ \bar{x}_0,\dots,\bar{x}_N}} \sum_{i=0}^{N-1} (\bar{x}_i - x_{ref})^\top Q(\bar{x}_i - x_{ref}) + \Delta\bar{u}_i^\top R \Delta\bar{u}_i + \tag{2.16a}$$

$$+ (\bar{x}_N - x_{ref})^\top Q_f (\bar{x}_N - x_{ref}),$$

$$\text{s.t.} \qquad \bar{x}_0 = x_t, \tag{2.16b}$$

$$\bar{x}_{k+1} = A\bar{x}_k + B\bar{u}_k, \ k = 0, \dots, N-1, \tag{2.16c}$$

$$\bar{u}_0 = u_{t-1} + \Delta\bar{u}_0, \tag{2.16d}$$

$$\bar{u}_k = \bar{u}_{k-1} + \Delta\bar{u}_k, \ k = 1, \dots N-1, \tag{2.16e}$$

$$\bar{x}_k \in \mathcal{X}_k, \ k = 0, \dots, N, \tag{2.16f}$$

$$\bar{u}_k \in \mathcal{U}_k, \ k = 0, \dots, N-1, \tag{2.16g}$$

$$\Delta\bar{u}_k \in \Delta\mathcal{U}_k, \ k = 0, \dots, N-1. \tag{2.16h}$$

The model in (2.16c) still uses the full control $\bar{u}$, which has to be determined from $\Delta\bar{u}$ and the previous $\bar{u}$ in (2.16e). In constraint (2.16d), the initial condition for $\bar{u}$ is set from the last control action applied u_{t-1}. Note that, in the presence of disturbances such that the previous control action u_{t-1} is not perfectly known, some static error can be introduced. Furthermore, this formulation is not necessarily a good one, since it does not penalize constant control actions, which is not desirable in systems with constrained energy/fuel. Reference [4] presents other strategies for introducing integral action.

Another method to track a state reference is with the use of a terminal constraint, as an optimization hard-constraint

$$\bar{x}_N = x_{ref}. \tag{2.17}$$

Because of the receding-horizon strategy, this constraint does not ensure that the system will reach the reference at sample N, or at all, since only the first optimal control action is applied and then the horizon slides forward. Thus, as the horizon N increases, the further away from the reference the steady-state system will be. To counteract this, the state cost matrices Q and Q_f can be set to zero, which allows for the system to converge to the state reference if the system has integral action, although not necessarily in N samples. For systems without integral action, the strategies in (2.15) and (2.16) can be used with the terminal constraint, in which case the advantage in using this constraint is to ensure system stability.

The terminal constraint can also be used to ensure the system reaches the reference in N samples, by decrementing the prediction horizon at each time sample, which is no longer the receding-horizon strategy. However, if the prediction horizon becomes short enough, the optimization problem can become infeasible, since it may be impossible to reach the reference from the initial condition in N samples, given the system dynamics, disturbances and the control and state constraints. This strategy is employed in Chap. 4 for the MPC controller for rendezvous.

2.2.2 State Substitution

In an MPC problem, the state at any time can be predicted solely from the initial condition and the sequence of control actions up to that time. This structure can be exploited to eliminate the state as an optimization variable, also eliminating the optimization constraints related to the prediction model, thus greatly simplifying the problem and allowing it to be solved faster, either with analytical or numerical optimization. Note that, however, numerical optimization algorithms specific for MPC can exploit its structure and often this technique is not applied.

Given the prediction model in matrix form as defined in Sect. 2.2

$$X = \tilde{A}X + \tilde{B}U + Ex(t), \tag{2.18}$$

where X and U are the state and control variables along the prediction horizon in vector form, and matrices $\tilde{A}$, $\tilde{B}$ and E are the augmented state matrices, as defined in (2.8). This model can be rewritten as

$$X = \underbrace{(I - \tilde{A})^{-1}\tilde{B}}_{F} U + \underbrace{(I - \tilde{A})^{-1}E}_{K} x(t). \tag{2.19}$$

Since matrix $\tilde{A}$ is lower triangular, the determinant of matrix $(I - \tilde{A})$ is 1 and thus this matrix is always invertible. It is also remarked that determining the inverse matrices in F and K may have a significant computational load for high values of the prediction horizon.

2.2.2.1 Quadratic Cost

For example, substituting X in the formulation with quadratic cost in (2.15) yields

$$\min_{U} \quad \begin{aligned} (FU + Kx(t) - X_{ref})^\top \tilde{Q}(FU + Kx(t) - X_{ref}) + \\ + (U - U_{ref})^\top \tilde{R}(U - U_{ref}), \end{aligned} \tag{2.20a}$$

$$\text{s.t.} \quad FU + Kx(t) \in \tilde{\mathcal{X}}, \tag{2.20b}$$

$$U \in \tilde{\mathcal{U}}. \tag{2.20c}$$

Simplifying and removing the terms independent from U, the cost function becomes

$$V(U) = \frac{1}{2}U^\top HU + \underbrace{\left(Jx(t) - LX_{ref} - \tilde{R}^\top U_{ref}\right)^\top}_{f} U, \tag{2.21}$$

with

$$H = F^\top \tilde{Q} F + \tilde{R},$$
$$J = F^\top \tilde{Q}^\top K, \tag{2.22}$$
$$L = F^\top \tilde{Q}^\top.$$

Notice that, by applying this substitution, the state has been eliminated as an optimization variable, as well as the constraints associated with the prediction model. Note, however, that the state is no longer directly accessible and for state constraints, it must be calculated from the control variables and initial condition, as can be seen in (2.20b).

2.2.3 ℓ_1-Norm Cost

Although less common than the quadratic cost, the ℓ_1-norm is also used for the MPC cost function instead of the quadratic cost (observe that the quadratic cost is the squared ℓ_2-norm). Denoting the ℓ_1-norm of a vector w by $\|w\|_1$, the cost functions become

$$l(x, u) = \|Q(x - x_{ref})\|_1 + \|Ru\|_1,$$
$$V_f(x) = \|Q_f(x - x_{ref})\|_1. \tag{2.23}$$

This cost function generates sparse solutions, in which the control becomes *bang–bang*, meaning that the actuators are either fully turned on or off. In matrix form, this formulation takes the shape of

$$\min_{X, U} \quad \|\tilde{Q}(X - X_{ref})\|_1 + \|\tilde{R}U\|_1, \tag{2.24a}$$

$$\text{s.t.} \quad X = \tilde{A}X + \tilde{B}U + \tilde{E}x(t), \tag{2.24b}$$

$$X \in \tilde{\mathcal{X}}, \tag{2.24c}$$

$$U \in \tilde{\mathcal{U}}. \tag{2.24d}$$

The drawback of the use of the ℓ_1 norm is the fact that the cost is no longer differentiable at the origin, and optimization algorithms require the use of non-smooth methods, using concepts like the sub-gradients and proximal operators [5].

2.2.4 LASSO Cost

Another possibility is to add an ℓ_1-norm control cost to the quadratic cost function, which is known as the LASSO cost function, being commonly used in regression analysis. The aim is to retain the desirable properties of the quadratic cost, such as robustness while adding some sparsity due to the ℓ_1-norm. In matrix form, the cost function becomes

$$V(X, U) = (X - X_{ref})^\top \tilde{Q}(X - X_{ref}) + U^\top \tilde{R}U + \|\tilde{R}_\lambda U\|_1, \qquad (2.25)$$

where R_λ is the control cost matrix associated with the ℓ_1-norm term.

A subclass of this formulation is to set $R = 0$, becoming similar to the formulation in Sect. 2.2.3 but with a quadratic cost on the state variables, therefore reducing sparsity regarding the tracking of the reference.

2.3 Nonlinear Model Predictive Control

Nonlinear MPC (NMPC) implies the control of systems with nonlinear models. Most real systems are nonlinear and sometimes these cannot be accurately approximated by a linearized model in the whole operating region. Thus, NMPC allows for the use of a nonlinear prediction model, which produces better state predictions, allowing for better control and to operate systems closer to the boundary of the admissible operating region [6]. On the other hand, NMPC incurs a higher computational cost, due to the necessity of solving an optimization problem with nonlinear and non-convex constraints.

The NMPC with quadratic cost is formulated as

$$\min_{\substack{\bar{u}_0,\dots,\bar{u}_{N-1} \\ \bar{x}_0,\dots,\bar{x}_N}} \sum_{i=0}^{N-1} \bar{x}_i^\top Q \bar{x}_i + \bar{u}_i^\top R \bar{u}_i + \bar{x}_N^\top Q_f \bar{x}_N, \qquad (2.26a)$$

$$\text{s.t.} \qquad \bar{x}_0 = x_t, \qquad (2.26b)$$

$$\bar{x}_{k+1} = f(\bar{x}_k, \bar{u}_k), \ k = 0, \dots, N-1, \qquad (2.26c)$$

$$\bar{x}_k \in \mathcal{X}_k, \ k = 0, \dots, N, \qquad (2.26d)$$

$$\bar{u}_k \in \mathcal{U}_k, \ k = 0, \dots, N-1, \qquad (2.26e)$$

where the constraint (2.26c) is now nonlinear.

For references on theory and implementation of NMPC, see the introduction in [6] and the tutorial [7].

2.4 Move Blocking

A strategy to reduce the complexity of the optimization problem, known as *move blocking*, is to shorten the number of control decisions. However, because decreasing the prediction horizon worsens the controller performance, a *control horizon* N_u is introduced, such that control actions beyond this horizon are all equal to the last control decision at time $k = N_u - 1$, as illustrated in Fig. 2.1, thus reducing the number of optimization variables. The control horizon is necessarily equal to or less than the prediction horizon N. The move-blocking formulation becomes

$$\min_{\substack{\bar{u}_0,\ldots,\bar{u}_{N_u-1} \\ \bar{x}_0,\ldots,\bar{x}_N}} \sum_{i=0}^{N_u-1} l_1(\bar{x}_i, \bar{u}_i) + \sum_{j=N_u}^{N-1} l_2(\bar{x}_j, u_{N_u-1}) + V_f(\bar{x}_N), \tag{2.27a}$$

$$\text{s.t.} \quad \bar{x}_0 = x_t, \tag{2.27b}$$

$$\bar{x}_{k+1} = f(\bar{x}_k, \bar{u}_k), \; k = 0, \ldots, N_u - 1, \tag{2.27c}$$

$$\bar{x}_{k+1} = f(\bar{x}_k, \bar{u}_{N_u-1}), \; k = N_u, \ldots, N - 1, \tag{2.27d}$$

$$\bar{x}_k \in \mathcal{X}_k, \; k = 0, \ldots, N, \tag{2.27e}$$

$$\bar{u}_k \in \mathcal{U}_k, \; k = 0, \ldots, N - 1. \tag{2.27f}$$

Because the number of control decisions is reduced, this complexity reduction strategy is suboptimal.

2.5 Experiments and Results

This section features MPC simulations of a simple linear system. The effect of using different cost functions is shown, as well as that of tuning its parameters, such as the prediction horizon and weight matrices, featuring both linear and nonlinear constraints. The aim of these experiments is not to obtain an optimal predictive controller for this system and with a feasible real-time implementation, but rather to show the capabilities and limitations of MPC. Furthermore, no disturbances are present in these simulations.

The following experiments are performed in an interpreted programming environment and running on a 4th Generation 2.4 GHz Intel-i7 Processor. In the absence of inequality constraints, the problems presented here can be readily solved analytically, except with an ℓ_1-norm cost on the state. Otherwise, numerical optimization algorithms are used to solve the optimal control problem. For the problems with a quadratic cost, linear model and linear constraints, the optimal control problem is a convex Quadratic Program (QP) and is solved with an interior-point algorithm [8]. For problems with nonlinear models or nonlinear constraints, it becomes a Nonlinear Program (NLP), for which a Sequential Quadratic Programming (SQP) algorithm [8] is used.

To solve each numerical optimization faster, the solution of one MPC iteration can, if the algorithm allows it, be used as the initial point for the next, a technique known as *warm start*. Furthermore, the state-substitution method presented in Sect. 2.2.2 is applied in order to reduce the number of optimization variables, which for the following experiments was empirically shown to decrease the computation time. Note, however, that some optimization algorithms may show better performance with a large but sparse problem, than with a smaller but denser one. Furthermore, some optimization algorithms specialized for MPC take advantage of the problem structure and do not apply this substitution, e.g. [9]. Although the average computation times of solving the MPC problem are presented for the following experiments, these may not be truly representative of a real-time implementation, since that is typically done on a compiled programming language. Therefore, the following computation times may be decreased.

2.5.1 Two-Dimensional Pure Inertial System

To experiment with linear MPC, a pure inertial system in two dimensions is considered. The system is described by two double integrators, and its discrete state model, if sampled with a zero-order hold (ZOH), is

$$
x_{k+1} = \underbrace{\begin{bmatrix} 1 & T_s & 0 & 0 \\ 0 & 1 & 0 & 0 \\ 0 & 0 & 1 & T_s \\ 0 & 0 & 0 & 1 \end{bmatrix}}_{A} x_k + \underbrace{\begin{bmatrix} T_s^2 & 0 \\ T_s & 0 \\ 0 & T_s^2 \\ 0 & T_s \end{bmatrix}}_{B} u_k, \tag{2.28}
$$

where T_s is the sampling period. In these experiments, the output model is not considered, since full state measurement is assumed. State variables x_1 and x_3 are the system position in the x-axis and y-axis, respectively, and states x_2 and x_4 are the corresponding velocities. The system has two inputs that allow it to move in any direction with no restrictions (holonomic movement). A sampling period of $T_s = 0.1$ s was used in the following experiments.

2.5.1.1 Unconstrained Problem

To begin with, no state or control constraints are considered and a quadratic cost function is used. Table 2.1 contains the controller parameters used in each experiment and the average time for solving the optimization problem (analytically).

In the first experiment, the system starts from the origin with no velocity, and the reference state is at coordinates (1, 2) with zero velocity. The prediction horizon is $N = 10$ samples, which together with the sampling period of 0.1 s grants a prediction of 1 s ahead, and all quadratic cost matrices are equal to the identity matrix. The results in Fig. 2.2 were obtained, where Fig. 2.2a contains the trajectory in the 2D space, and Fig. 2.2b present the state (top) and control variables (bottom) as a function of time. Note that the position and velocity vectors for each direction are plotted with different

Table 2.1 Controller parameters and computation times for unconstrained MPC experiments

Figures	N	R	Q	Q_f	t_{avg} (µs)
2.2	10	I	I	I	50
2.3	10	$5I$	I	I	48
2.4	20	$5I$	I	I	71

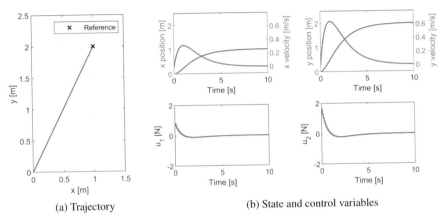

(a) Trajectory   (b) State and control variables

Fig. 2.2 Control of pure inertial system with no constraints

scales. The system goes in a straight line toward the reference, requiring more control action from the second input to do so since the reference in the y-axis is further away. The control values are constant in each time sample, due to the ZOH discretization.

In the next experiment, the weight of the control action in the cost function is increased. It can be observed in Fig. 2.3 that the control action is now less energetic, and as a result, the system overshoots the reference and takes longer to converge. This is due to the fact that the control action now weighs more in the cost function, and so the minimum is such that there is less control action and a greater tracking error. On the other hand, increasing cost matrices Q and Q_f relative to R results in a more aggressive control action in order to track the reference more closely. The cost matrices can then be used to tune the trade-off between energy spent controlling the system and the speed at which it converges to the reference.

Lastly, the prediction horizon is doubled, and the results in Fig. 2.4 are obtained. The control action is again more aggressive, and the reference tracking is better and without overshoot, an effect which is the same as that of decreasing the control cost. In fact, in the absence of any constraints, changing the prediction horizon has a similar effect as tuning the cost matrices. However, the prediction horizon also has an effect on system stability [2], and if the prediction horizon is too short, the closed-loop system can become unstable. With the increase of the prediction horizon N, the state is predicted and optimized for longer into the future, which increases stability. In the limit, an infinite prediction horizon guarantees an asymptotically

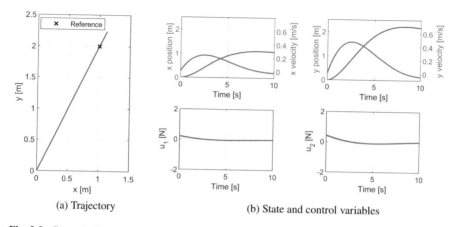

(a) Trajectory

(b) State and control variables

Fig. 2.3 Control of pure inertial system with no constraints and increased control cost

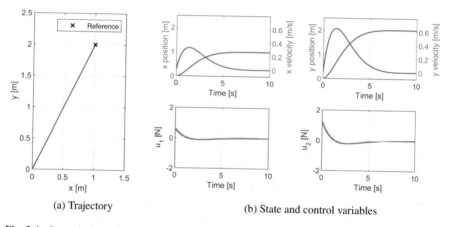

(a) Trajectory

(b) State and control variables

Fig. 2.4 Control of pure inertial system with no constraints and increased prediction horizon

stable closed-loop system, if the open-loop system is stabilizable [1]. Furthermore, it can be seen from Table 2.1 that the computation time increases with the prediction horizon, as expected.

Another important consideration is that increasing N also increases the number of optimization variables, and thus increases the computation time for solving the optimization problem. As can be seen from Table 2.1, these optimal control problems can be solved very efficiently, which is due to the fact that the problem is unconstrained and can be solved analytically. In the presence of inequality constraints, however, numerical optimization algorithms are required.

Table 2.2 Controller parameters and computation times for constrained MPC experiments

Figures	N	R	Q	Q_f	Constraints	t_{avg} (ms)
2.5	10	I	$10I$	$10I$	Control bounds	1.5
2.6	10	I	$10I$	$10I$	Control bounds, circular obstacle	18
2.7	20	I	$10I$	$10I$	Control bounds, circular obstacle	41
2.8	10	I	$10I$	$10I$	Control bounds, 4 circular obstacles	26
2.9	10	I	$10I$	$10I$	Control bounds, square obstacle	21

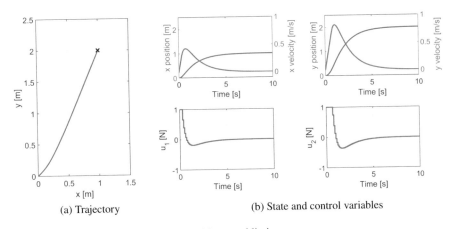

(a) Trajectory

(b) State and control variables

Fig. 2.5 Control of pure inertial system with control limits

2.5.1.2 Control and State Constraints

Applying lower and upper bounds of ± 1 N on the control variables with optimization constraints, the result in Fig. 2.5 is obtained. The controller parameters used are contained in Table 2.2, as well as the average computation times. The system now converges asymmetrically towards the reference since a straight-line trajectory would require more control action from u_2, and both control variables are saturated at the beginning.

An obstacle avoidance constraint is now added, which constrains the position state variables. As seen in Fig. 2.6, the system goes around the obstacle but, because it is modelled as a single point object, it goes very close and along the obstacle border since it is the most efficient trajectory that satisfies the constraints. It can also be observed that, initially, the system travels towards the object, in a similar trajectory as in Fig. 2.5, and, as the prediction horizon reaches it, the system starts to swerve.

Increasing the prediction horizon, it can be observed in Fig. 2.7 that the system now diverts its trajectory earlier to avoid the obstacle, since it is detected earlier.

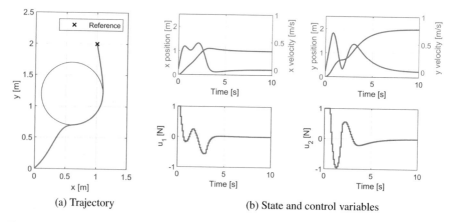

(a) Trajectory   (b) State and control variables

Fig. 2.6 Control of pure inertial system with a circular obstacle

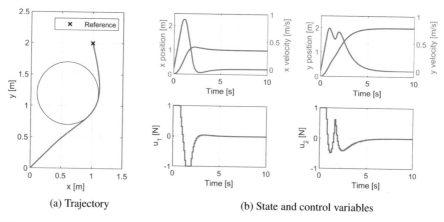

(a) Trajectory   (b) State and control variables

Fig. 2.7 Control of pure inertial system with a circular obstacle and increased prediction horizon

Moreover, the control sequence is now smoother, with fewer corrections along the way, because the controller has a better plan of the trajectory.

Adding multiple circular obstacles, the results in Fig. 2.8 are obtained. Note that obstacle avoidance constraints are non-convex, meaning that the optimization problem may have several local minima to which the algorithm can converge, depending on the initial point. For example, any solution for which the trajectory goes around a different side of an obstacle is a local minimum. While the trajectory in Fig. 2.8 appears to be a global minimizer, it might not be, depending on the cost function. This particular solution requires several direction corrections to avoid the obstacles, and so, for some cost weighting the global minimum might be to go around all the obstacles. Globally optimizing non-convex problems is usually performed by solving the problem several times with different initial points, which can be infeasible to do in real time.

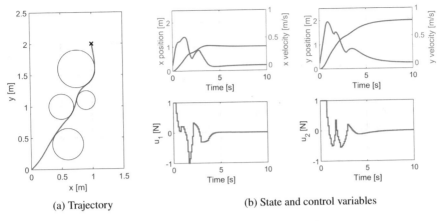

(a) Trajectory

(b) State and control variables

Fig. 2.8 Control of pure inertial system with multiple circular obstacles

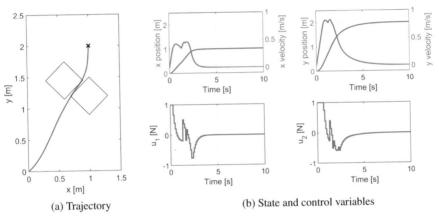

(a) Trajectory

(b) State and control variables

Fig. 2.9 Control of pure inertial system with two square obstacles

Figure 2.9 shows an experiment with two square obstacles, where it can be observed that the trajectory passes through the corners of the obstacle. This is a limitation of the way the obstacle constraint is formulated, since only the discrete points are constrained, and not the whole continuous trajectory. Upon inspecting the discrete positions along the trajectories, it is observed that the constraints are in fact satisfied. It is possible to formulate an obstacle constraint that restricts the line between discrete points, at the cost of extra computing power. Another way to minimize this effect is to decrease the sampling period or to use intermediate samples for the constraints only. It is also possible to constraint the state in continuous time via semidefinite programming [10].

2.5.1.3 ℓ_1-Norm Cost

Using the ℓ_1-norm for the cost function in the absence of constraints yields the result in Fig. 2.10, with the controller parameters presented in Table 2.3. The cost on the velocity states has to be reduced, otherwise, these would weigh too much on the cost function and the controller would generate no action, due to the sparsity of the ℓ_1-norm. It can be seen that the actuators are turned on for one sample only, resting afterwards, resulting in a constant velocity trajectory. As the system approaches the reference, its velocity is cancelled, and thus the control is *bang–bang*.

Adding control limits in Fig. 2.11, the acceleration and deceleration are no longer performed in one sample, due to control saturation, and the system converges asymmetrically to the reference.

2.5.1.4 LASSO Cost

Using the LASSO cost yields the results in Fig. 2.12, using the parameters in Table 2.4. At the start, the control is sparse, resembling that obtained with the ℓ_1-norm cost, and afterwards, it is more active, like the control obtained with the quadratic cost.

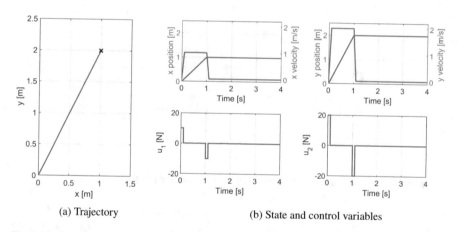

(a) Trajectory (b) State and control variables

Fig. 2.10 Control of pure inertial system with ℓ_1-norm cost and no constraints

Table 2.3 Controller parameters and computation times for ℓ_1-norm MPC experiments

Figures	N	R	Q	Q_f	Constraints	t_{avg} (ms)
2.10	10	I	10diag(1, 0, 1, 0)	100diag(1, 0, 1, 0)	None	179
2.11	10	I	10diag(1, 0, 1, 0)	100diag(1, 0, 1, 0)	Control bounds	281

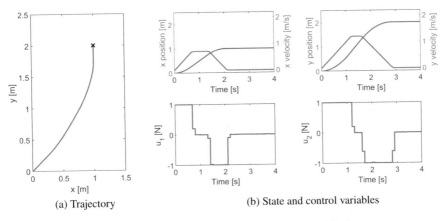

(a) Trajectory

(b) State and control variables

Fig. 2.11 Control of pure inertial system with ℓ_1-norm cost and control limits

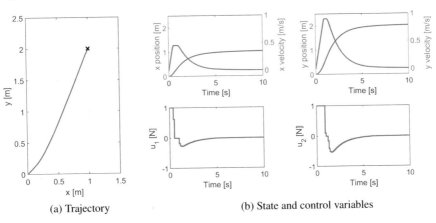

(a) Trajectory

(b) State and control variables

Fig. 2.12 Control of pure inertial system with LASSO cost

Table 2.4 Controller parameters and computation times for LASSO cost MPC experiments

Figures	N	R	R_λ	Q	Q_f	Constraints	t_{avg} (ms)
2.12	10	I	I	$50I$	$100I$	Control bounds	4.17
2.13	10	0	I	$50I$	$100I$	Control bounds	3.98

In Fig. 2.13, the quadratic cost on the control variables is removed entirely, by setting $R = 0$. The control action is now less smooth at the beginning of the simulation.

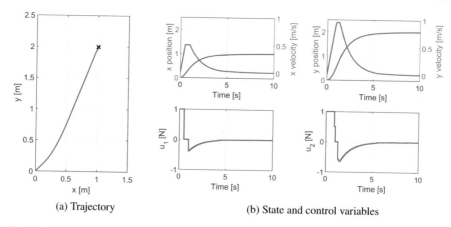

(a) Trajectory   (b) State and control variables

Fig. 2.13 Control of pure inertial system with LASSO cost and zero quadratic cost for control variables

References

1. J. Rawlings, D. Mayne, M. Diehl, *Model Predictive Control: Theory, Computation, and Design*, 2nd edn. (Nob Hill Publishing, 2017)
2. D.Q. Mayne, J.B. Rawlings, C.V. Rao, P.O. Scokaert, Constrained model predictive control: stability and optimality. Automatica **36**, 789–814 (2000)
3. G.F. Franklin, J. Powell, M. Workman, *Digital Control of Dynamic Systems*, 3rd edn. (Addison-Wesley, 2006)
4. D. Ruscio, Model predictive control with integral action: a simple MPC algorithm. Model. Identif. Control: Nor. Res. Bull. **34**, 119–129 (2013)
5. M. Nagahara, *Sparsity Methods for System and Contram* (NowPublishers Incorporated, 2020)
6. R. Findeisen, F. Allgöwer, *An Introduction to Nonlinear Model Predictive Control* (2002)
7. J.B. Rawlings, Tutorial: model predictive control technology in *Proceedings of the 1999 American Control Conference (Cat. No. 99CH36251)*, vol. 1, June 1999 (1999), pp. 662–676
8. J. Nocedal, S. Wright, *Numerical Optimization* (Springer Science & Business Media, 2006)
9. Y. Wang, S. Boyd, Fast model predictive control using online optimization. IEEE Trans. Control. Syst. Technol. **18**, 267–278 (2009)
10. G. Deaconu, C. Louembet, A. Théron, Designing continuously constrained spacecraft relative trajectories for proximity operations. J. Guid. Control Dyn. **38**, 1208–1217 (2014)

Chapter 3
Relative Orbital Mechanics

Relative orbital mechanics refers to the relative motion of two satellites orbiting the same body. In an orbital rendezvous context, when the two spacecraft are in close range, it is convenient to consider relative positions and velocities, centred at one of the spacecraft, rather than absolute coordinates centred in the central body.

The equations that describe the motion of an orbiting satellite and the relative motion of two satellites are derived from Newton's law of gravitation and from his second law of motion. These result in nonlinear differential equations, that can be infeasible to solve and work within real-time applications, such as rendezvous trajectory guidance. Hence, it is possible to approximate the nonlinear equations for the relative motion and maintain accuracy if the spacecraft are sufficiently close. For the special case of a circular orbit, these approximations result in the well-known Hill equations [1], which describe a dynamical system that is linear and time-invariant. For the general case of an elliptic orbit, the approximation yields the Yamanaka–Ankersen state transition matrix [2], that describes a discrete linear time-variant (LTV) system.

As discussed in Chap. 1, there are many different linearized relative dynamic models available in the literature. These may include, for example, perturbations such as J_2 [3] and atmospheric drag [4]. Other models are based on relative state representations different than Cartesian coordinates, for example, Relative Orbital Elements (ROEs) [5–7]. Many other models with different assumptions, applications and complexity are also available [8, 9]. Since, however, the focus of this book is not the accurate modelling of relative dynamics, we limit ourselves to the Yamanaka–Ankersen model, which includes no perturbations and is not necessarily the most advanced model available, but which is the state-of-the-art baseline solution for linearised relative dynamics. Furthermore, this implies no loss of generality for the application of the rendezvous methods presented in this book, which is its real focus, since these are very versatile to whatever dynamic model is chosen, as will be discussed in Chap. 4.

In this chapter, we first introduce the nonlinear dynamics on the inertial frame in Sect. 3.1. In Sect. 3.2, a non-inertial frame of reference centred on one of the satellites is presented, and in Sect. 3.3, the approximated dynamics in this frame of reference

© The Author(s), under exclusive license to Springer Nature Switzerland AG 2021
A. Botelho et al., *Predictive Control for Spacecraft Rendezvous*,
SpringerBriefs in Applied Sciences and Technology,
https://doi.org/10.1007/978-3-030-75696-3_3

are derived. Sections 3.4 and 3.5 present several simulations of free-drift motions and thrust manoeuvres for circular and elliptic target orbits, respectively.

3.1 Nonlinear Inertial Dynamics

Consider two satellites, treated as point masses, in motion around a central body, such that their gravitational pull on each other is negligible. In a *rendezvous* context, one of the satellites is in free motion, designated *target* spacecraft, while the other, the *chaser*, performs manoeuvres to close their relative positions. Define the auxiliary function

$$\mathbf{f}_g(\mathbf{r}) = -\frac{\mu}{r^3}\mathbf{r}, \tag{3.1}$$

where $\mathbf{r}$ is a position vector, r its magnitude and μ is the standard gravitational parameter. The motion of the target and chaser spacecraft, with positions $\mathbf{r}_t$ and $\mathbf{r}_c$ relative to the inertial frame of reference, is then given by

$$\ddot{\mathbf{r}}_t = \mathbf{f}_g(\mathbf{r}_t), \tag{3.2}$$

$$\ddot{\mathbf{r}}_c = \mathbf{f}_g(\mathbf{r}_c) + \frac{\mathbf{F}}{m_c}, \tag{3.3}$$

where $\mathbf{F}$ is the force vector applied by the chaser actuators and m_c is the mass of the chaser.

The relative position between the two satellites, illustrated in Fig. 3.1, is defined as $\mathbf{s} = \mathbf{r}_c - \mathbf{r}_t$, and so the relative motion satisfies

$$\ddot{\mathbf{s}} = \mathbf{f}_g(\mathbf{r}_c) - \mathbf{f}_g(\mathbf{r}_t) + \frac{\mathbf{F}}{m_c}. \tag{3.4}$$

Unlike in the case of only one unperturbed satellite, this problem has no closed-form solution and must be solved numerically or approximated with a linearization.

3.2 Target Local Orbital Frame

When the distance between the two spacecraft is short, it is convenient to consider the non-inertial *target local orbital frame*, illustrated in Fig. 3.2, also known as *local-vertical/local-horizontal frame* (LVLH), centred in the target spacecraft. The axis x_{lo} is in the general direction of the velocity vector, although it is not always aligned with it, and is commonly known as V-bar. Axis y_{lo} is orthogonal to the orbital plane,

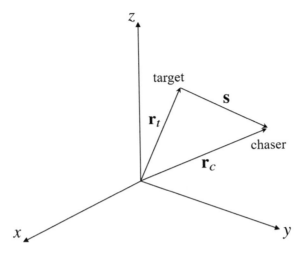

Fig. 3.1 Relative position of target and chaser spacecraft on the inertial frame

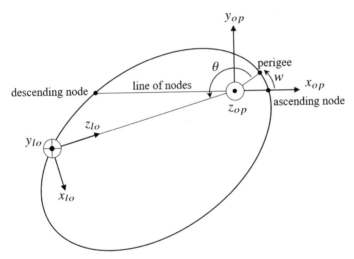

Fig. 3.2 The target local orbital frame (F_{lo})

in the opposite direction of the angular momentum, and is also known as H-bar. The axis z_{lo}, known as R-bar, is always directed to the centre of mass of the central body. For this reason, the frame rotates with the orbital angular velocity ω and thus is non-inertial.

To determine coordinates in this frame (F_{lo}) from ones in the inertial orbital plane frame (F_{op}) [10], the target position $\mathbf{r}_t$ is first subtracted, and then a counter-clockwise rotation around z_{op} by the argument of perigee w and the true anomaly θ is performed. Then, another counter-clockwise rotation around the z-axis by 90° is applied, followed by a clockwise rotation around the x-axis by 90°. The coordinate transformation is then

$$\begin{bmatrix} x_{lo} \\ y_{lo} \\ z_{lo} \end{bmatrix} = \begin{bmatrix} 1 & 0 & 0 \\ 0 & 0 & -1 \\ 0 & 1 & 0 \end{bmatrix} \begin{bmatrix} 0 & 1 & 0 \\ -1 & 0 & 0 \\ 0 & 0 & 1 \end{bmatrix} \begin{bmatrix} \cos\alpha & \sin\alpha & 0 \\ -\sin\alpha & \cos\alpha & 0 \\ 0 & 0 & 1 \end{bmatrix} \begin{bmatrix} x_{op} - x_t \\ y_{op} - y_t \\ z_{op} - z_t \end{bmatrix}, \qquad (3.5)$$

with $\alpha = \theta + w$. The velocity vectors in an inertial frame and a rotating (*) frame with angular velocity vector $\boldsymbol{\omega}$ are related by

$$\frac{d^* \mathbf{s}^*}{dt} = -\boldsymbol{\omega} \times \mathbf{s}^* + \frac{d\mathbf{s}}{dt}. \qquad (3.6)$$

This reference frame is generally used in a rendezvous context in order to represent the chaser spacecraft position and velocity relative to the target [10].

3.3 Approximate Equations of Relative Motion

As shown in Fehse [10], applying a first-order Taylor expansion to $\mathbf{f}_g(\mathbf{r}_c)$ around the target position $\mathbf{r}_t$ yields

$$\mathbf{f}_g(\mathbf{r}_c) \approx \mathbf{f}_g(\mathbf{r}_t) + \frac{d\mathbf{f}_g(\mathbf{r})}{d\mathbf{r}}\bigg|_{\mathbf{r}=\mathbf{r}_t} (\mathbf{r}_c - \mathbf{r}_t). \qquad (3.7)$$

Applying (3.7) to the equation of the chaser motion in (3.3) and given (3.2), the relative motion is approximated by

$$\ddot{\mathbf{s}} = \frac{d\mathbf{f}_g(\mathbf{r})}{d\mathbf{r}}\bigg|_{\mathbf{r}=\mathbf{r}_t} \mathbf{s} + \frac{\mathbf{F}}{m_c}. \qquad (3.8)$$

Differentiating (3.6), the relation between the acceleration in the inertial and rotating (*) frames is

$$\frac{d^2\mathbf{s}}{dt^2} = \frac{d^{*2}\mathbf{s}^*}{dt^2} + \boldsymbol{\omega} \times (\boldsymbol{\omega} \times \mathbf{s}^*) + 2\boldsymbol{\omega} \times \frac{d^*\mathbf{s}^*}{dt} + \frac{d\boldsymbol{\omega}}{dt} \times \mathbf{s}^*, \qquad (3.9)$$

where the last three terms are the centrifugal, Coriolis and Euler fictitious forces, due to the expression of the acceleration in a rotating frame.

Substituting (3.9) in Eq. (3.8), we get

$$\frac{d^{*2}\mathbf{s}^*}{dt^2} + \boldsymbol{\omega} \times (\boldsymbol{\omega} \times \mathbf{s}^*) + 2\boldsymbol{\omega} \times \frac{d^*\mathbf{s}^*}{dt} + \frac{d\boldsymbol{\omega}}{dt} \times \mathbf{s}^* - \frac{d\mathbf{f}_g(\mathbf{r})}{d\mathbf{r}}\bigg|_{\mathbf{r}=\mathbf{r}_t} \mathbf{s}^* = \frac{\mathbf{F}}{m_c}. \qquad (3.10)$$

As shown in [10], after computing all the cross products and the Jacobian, the simplification of (3.10) for the general case of an elliptical orbit is

$$\ddot{x} - \omega^2 x - 2\omega\dot{z} - \dot{\omega}z + k\omega^{\frac{3}{2}}x = \frac{F_x}{m_c}, \tag{3.11a}$$

$$\ddot{y} + k\omega^{\frac{3}{2}}y = \frac{F_y}{m_c}, \tag{3.11b}$$

$$\ddot{z} - \omega^2 z + 2\omega\dot{x} + \dot{\omega}x - 2k\omega^{\frac{3}{2}}z = \frac{F_z}{m_c}, \tag{3.11c}$$

where $\mathbf{s}^* = [x, y, z]^\top$, $\mathbf{F} = \begin{bmatrix} F_x, F_y, F_z \end{bmatrix}^\top$ and k is the constant $k = \mu/h^{\frac{3}{2}}$, with h as the magnitude of the target orbit-specific angular momentum. These are known as the Linearized Equations of Relative Motion (LERM).

The set of differential equations in (3.11) is linear with respect to the relative position, velocity and acceleration, although it is not so with respect to the angular velocity ω, which is not constant in the case of non-circular orbits. This results in the relative motion dynamics being time-variant.

Notice also that the out-of-plane motion (H-bar) in (3.11b) has become detached from the in-plane motion (V-bar and R-bar), which simplifies the problem since the two can be solved separately. This is a result of the linearisation, and for the nonlinear dynamics, the two motions are in fact coupled.

These equations are the result of linear approximations, with respect to position, of the real nonlinear motion, and are only accurate if the distance between the target and the centre of mass of the central body is significantly greater than the distance between the target and chaser spacecraft.

3.3.1 Circular Orbit Case

For the special case of a circular target orbit, the orbital angular velocity is constant $\omega = \mu/r_t^3$ and so $\dot{\omega} = 0$. Since $h = \omega r^2$, we then get $k = w^{1/2}$. Substituting, Eqs. (3.11a-c) simplify to

$$\ddot{x} - 2\omega\dot{z} = \frac{F_x}{m_c}, \tag{3.12a}$$

$$\ddot{y} + \omega^2 y = \frac{F_y}{m_c}, \tag{3.12b}$$

$$\ddot{z} + 2\omega\dot{x} - 3\omega^2 z = \frac{F_z}{m_c}. \tag{3.12c}$$

This system of linear differential equations is known as the Hill equations [1], and also sometimes as the Hill–Clohessy–Wiltshire equations [11]. They describe a linear and time-invariant dynamical system, that can be represented in state space with the model

$$
\begin{bmatrix} \dot{x} \\ \dot{z} \\ \ddot{x} \\ \ddot{z} \end{bmatrix} = \begin{bmatrix} 0 & 0 & 1 & 0 \\ 0 & 0 & 0 & 1 \\ 0 & 0 & 0 & 2\omega \\ 0 & 3\omega^2 & -2\omega & 0 \end{bmatrix} \begin{bmatrix} x \\ z \\ \dot{x} \\ \dot{z} \end{bmatrix} + \begin{bmatrix} 0 & 0 \\ 0 & 0 \\ \frac{1}{m_c} & 0 \\ 0 & \frac{1}{m_c} \end{bmatrix} \begin{bmatrix} F_x \\ F_z \end{bmatrix} , \tag{3.13}
$$

for the in-plane dynamics, and

$$
\begin{bmatrix} \dot{y} \\ \ddot{y} \end{bmatrix} = \begin{bmatrix} 0 & 1 \\ \omega^2 & 0 \end{bmatrix} \begin{bmatrix} y \\ \dot{y} \end{bmatrix} + \begin{bmatrix} 0 \\ \frac{1}{m_c} \end{bmatrix} [F_y] , \tag{3.14}
$$

for the out-of-plane motion. The closed-form solution for the Hill equations, known as the Clohessy–Wiltshire equations, is trivial and can be found along with their derivation in [10].

3.3.2 Simplification of General Equations

The approximate equations of relative motion in (3.11) may be simplified in order to help find a solution, by changing the independent variable from time to the true anomaly (i.e. the angle that defines the position of the body moving along the orbit), and by applying a coordinate change. With the chain rule, the time derivative of a variable r is related to its derivative in respect to the true anomaly θ by

$$
\frac{dr}{dt} = \frac{dr}{d\theta} \frac{d\theta}{dt} . \tag{3.15}
$$

Given that the time derivative of the true anomaly θ is the orbital angular velocity ω, and denoting the derivative in respect to θ by r', the above expression simplifies to

$$
\dot{r} = \omega r' , \tag{3.16}
$$

and the second time derivative is

$$
\ddot{r} = \omega^2 r'' + \omega \omega' r' . \tag{3.17}
$$

Applying the derivatives (3.16) and (3.17) to the positions and the angular velocity in (3.11) yields

$$\omega^2 x'' + \omega\omega' x' + (k\omega^{\frac{3}{2}} - \omega^2)x - 2\omega^2 z' - \omega\omega' z = \frac{F_x}{m_c}, \qquad (3.18a)$$

$$\omega^2 y'' + \omega\omega' y' + k\omega^{\frac{3}{2}} y = \frac{F_y}{m_c}, \qquad (3.18b)$$

$$\omega^2 z'' + \omega\omega' z' - (2k\omega^{\frac{3}{2}} + \omega^2)z + 2\omega^2 x' + \omega\omega' x = \frac{F_z}{m_c}. \qquad (3.18c)$$

The orbital angular velocity can be rewritten as

$$\omega = (1 + e\cos\theta)^2 \frac{\mu^2}{h^3}. \qquad (3.19)$$

Defining the auxiliary parameter

$$\rho(\theta) = 1 + e\cos(\theta), \qquad (3.20)$$

and since $k = \mu/h^{3/2}$, the angular velocity becomes

$$\omega = k^2 \rho^2, \qquad (3.21)$$

and its derivative with respect to θ is

$$\omega' = -2k^2 \rho e \sin(\theta). \qquad (3.22)$$

Substituting with (3.21) and (3.22), (3.18) becomes

$$\rho x'' - 2e\sin\theta x' - e\cos\theta x - 2\rho z' + 2e\sin\theta z = \frac{F_x}{m_c k^4 \rho^3}, \qquad (3.23a)$$

$$\rho y'' - 2e\sin\theta y' + y = \frac{F_y}{m_c k^4 \rho^3}, \qquad (3.23b)$$

$$\rho z'' - 2\sin\theta z' - (3 + e\cos\theta)z + 2\rho x' - 2e\sin\theta x = \frac{F_z}{m_c k^4 \rho^3}. \qquad (3.23c)$$

Applying the coordinate transformation,

$$\begin{bmatrix} \tilde{x} \\ \tilde{y} \\ \tilde{z} \end{bmatrix} = \rho(\theta) \begin{bmatrix} x \\ y \\ z \end{bmatrix}, \qquad (3.24)$$

the derivatives become

$$\begin{bmatrix} \tilde{x}' \\ \tilde{y}' \\ \tilde{z}' \end{bmatrix} = \rho(\theta) \begin{bmatrix} x' \\ y' \\ z' \end{bmatrix} - e\sin\theta \begin{bmatrix} x \\ y \\ z \end{bmatrix}, \qquad (3.25)$$

and the second derivatives

$$
\begin{bmatrix} \tilde{x}'' \\ \tilde{y}'' \\ \tilde{z}'' \end{bmatrix} = \rho(\theta) \begin{bmatrix} x'' \\ y'' \\ z'' \end{bmatrix} - 2e \sin \theta \begin{bmatrix} x' \\ y' \\ z' \end{bmatrix} - e \cos \theta \begin{bmatrix} x \\ y \\ z \end{bmatrix} . \tag{3.26}
$$

Writing (3.23) as a function of the second derivatives, substituting them in (3.26) and applying the transformations (3.24) and (3.25) yields the simplified equations of relative motion in the domain of θ

$$
\tilde{x}'' - 2\tilde{z}' = \frac{F_x}{m_c k^4 \rho^3}, \tag{3.27a}
$$

$$
\tilde{y}'' + \tilde{y} = \frac{F_y}{m_c k^4 \rho^3}, \tag{3.27b}
$$

$$
\tilde{z}'' - \frac{3}{\rho}\tilde{z} + 2\tilde{x}' = \frac{F_x}{m_c k^4 \rho^3} . \tag{3.27c}
$$

These are known as the Tschauner–Hempel equations [12] (also sometimes known as the Lawden equations [13]), and are an easier set of ordinary differential equations to solve than (3.11).

3.3.3   Homogeneous Solution

A simple homogeneous solution ($\mathbf{F} = 0$) to (3.27), in the form of a state transition matrix, was introduced by [2], and further detailed by [14]. The transition matrix propagates the state in the domain of the true anomaly from an initial θ_0, at time t_0, to the state at θ_t, for time t.

First, given initial conditions on position and velocity at time t_0, the transformed position and velocities must be determined for $\theta = \theta_0$ with (3.24) for the position, and for the velocity,

$$
\begin{bmatrix} \tilde{x}' \\ \tilde{y}' \\ \tilde{z}' \end{bmatrix} = -e \sin \theta \begin{bmatrix} x \\ y \\ z \end{bmatrix} + \frac{1}{k^2 \rho(\theta)} \begin{bmatrix} \dot{x} \\ \dot{y} \\ \dot{z} \end{bmatrix} . \tag{3.28}
$$

In matrix form, the transformations become

$$
\begin{bmatrix} \tilde{x} \\ \tilde{z} \\ \tilde{x}' \\ \tilde{z}' \end{bmatrix} = \underbrace{\begin{bmatrix} \rho(\theta) & 0 & 0 & 0 \\ 0 & \rho(\theta) & 0 & 0 \\ -e\sin\theta & 0 & \frac{1}{k^2\rho(\theta)} & 0 \\ 0 & -e\sin\theta & 0 & \frac{1}{k^2\rho(\theta)} \end{bmatrix}}_{\Lambda_i(\theta)} \begin{bmatrix} x \\ z \\ \dot{x} \\ \dot{z} \end{bmatrix}, \quad \begin{bmatrix} \tilde{y} \\ \tilde{y}' \end{bmatrix} = \underbrace{\begin{bmatrix} \rho(\theta) & 0 \\ -e\sin\theta & \frac{1}{k^2\rho(\theta)} \end{bmatrix}}_{\Lambda_o(\theta)} \begin{bmatrix} y \\ \dot{y} \end{bmatrix}.
$$

$$(3.29)$$

Afterwards, the so-called pseudo-initial conditions must be computed for the in-plane motion with

$$
\begin{bmatrix} \bar{x}_0 \\ \bar{z}_0 \\ \bar{x}_0' \\ \bar{z}_0' \end{bmatrix} = \frac{1}{1-e^2} \underbrace{\begin{bmatrix} 1-e^2 & 3es(1/\rho+1/\rho^2) & -es(1+1/\rho) & -ec+2 \\ 0 & -3s(1/\rho+e^2/\rho^2) & s(1+1/\rho) & c-2e \\ 0 & -3(c/p+e) & c(1+1/\rho)+e & -s \\ 0 & 3\rho+e^2-1 & -\rho^2 & es \end{bmatrix}}_{\phi_i^{-1}(\theta_0)}_{\theta_0} \begin{bmatrix} \tilde{x}_0 \\ \tilde{z}_0 \\ \tilde{x}_0' \\ \tilde{z}_0' \end{bmatrix},
$$

$$(3.30)$$

with $s(\theta) = \rho\sin\theta$ and $c(\theta) = \rho\cos\theta$. Note that, for simplification, dependencies on θ for ρ, s and c were omitted. However these parameters must be computed for $\theta = \theta_0$. For the out-of-plane motion, no pseudo-initial conditions are computed.

The state at a time t with true anomaly θ_t can be computed from the state at time t_0 with a transition matrix, and so we have for the in-plane motion

$$
\begin{bmatrix} \tilde{x}_t \\ \tilde{z}_t \\ \tilde{x}_t' \\ \tilde{z}_t' \end{bmatrix} = \underbrace{\begin{bmatrix} 1 & -c(1+1/\rho) & s(1+1/\rho) & 3\rho^2 J \\ 0 & s & c & (2-3esJ) \\ 0 & 2s & 2c-e & 3(1-2esJ) \\ 0 & s' & c' & -3e(s'J+s/\rho^2) \end{bmatrix}}_{\phi_i(\theta_t)}_{\theta_t} \begin{bmatrix} \bar{x}_0 \\ \bar{z}_0 \\ \bar{x}_0' \\ \bar{z}_0' \end{bmatrix}, \quad (3.31)
$$

and for the out-of-plane motion

$$
\begin{bmatrix} \tilde{y}_t \\ \tilde{y}_t' \end{bmatrix} = \underbrace{\begin{bmatrix} \cos(\theta_t-\theta_0) & \sin(\theta_t-\theta_0) \\ -\sin(\theta_t-\theta_0) & \cos(\theta_t-\theta_0) \end{bmatrix}}_{\phi_o(\theta_0,\theta_t)} \begin{bmatrix} \tilde{y}_0 \\ \tilde{y}_0' \end{bmatrix}, \quad (3.32)
$$

with $s' = \cos\theta + e\cos 2\theta$, $c' = -(\sin\theta + e\sin 2\theta)$ and $J = k^2(t-t_0)$.

The transformed position and velocity at time t must then be reverted. The inverse transformation for the position is then

$$
\begin{bmatrix} x \\ y \\ z \end{bmatrix} = \frac{1}{\rho(\theta)} \begin{bmatrix} \tilde{x} \\ \tilde{y} \\ \tilde{z} \end{bmatrix}, \quad (3.33)
$$

and for the velocity

$$\begin{bmatrix} \dot{x} \\ \dot{y} \\ \dot{z} \end{bmatrix} = k^2 e \sin \theta \begin{bmatrix} \tilde{x} \\ \tilde{y} \\ \tilde{z} \end{bmatrix} + k^2 \rho(\theta) \begin{bmatrix} \tilde{x}' \\ \tilde{y}' \\ \tilde{z}' \end{bmatrix} . \tag{3.34}$$

In matrix form, the transformation is

$$\begin{bmatrix} x \\ z \\ \dot{x} \\ \dot{z} \end{bmatrix} = \underbrace{\begin{bmatrix} \frac{1}{\rho(\theta)} & 0 & 0 & 0 \\ 0 & \frac{1}{\rho(\theta)} & 0 & 0 \\ k^2 e \sin \theta & 0 & k^2 \rho(\theta) & 0 \\ 0 & k^2 e \sin \theta & 0 & k^2 \rho(\theta) \end{bmatrix}}_{\Lambda_i^{-1}(\theta)} \begin{bmatrix} \tilde{x} \\ \tilde{z} \\ \tilde{x}' \\ \tilde{z}' \end{bmatrix} , \quad \begin{bmatrix} y \\ \dot{y} \end{bmatrix} = \underbrace{\begin{bmatrix} \frac{1}{\rho(\theta)} & 0 \\ k^2 e \sin \theta & k^2 \rho(\theta) \end{bmatrix}}_{\Lambda_o^{-1}(\theta)} \begin{bmatrix} \tilde{y} \\ \tilde{y}' \end{bmatrix} . \tag{3.35}$$

The true anomaly at time t can be computed from t_0 and θ_0 by first computing the eccentric anomaly E at time t_0 with

$$E = \arctan 2 \left(\sqrt{1 - e^2} \sin \theta, \ e + \cos \theta \right), \tag{3.36}$$

and then calculating the mean anomaly M at time t_0 with Kepler's equation

$$M = E - e \sin E. \tag{3.37}$$

The mean anomaly at time t can be determined with

$$M_t = M_0 + \frac{2\pi}{T}(t - t_0), \tag{3.38}$$

where T is the target orbital period, and the eccentric anomaly E_t can then be obtained by solving Kepler's equation (3.37) w.r.t the eccentric anomaly E. Finally, the true anomaly θ_t is computed with

$$\theta = 2 \arctan \left(\sqrt{\frac{1 + e}{1 - e}} \tan \frac{E}{2} \right). \tag{3.39}$$

Summarizing, to determine the state at time t from the state at time t_0 one must

- Compute the transformed position and velocity with (3.24) and (3.28) for $\theta = \theta_0$;
- Compute the pseudo-initial conditions for the in-plane motion with (3.30);
- Compute true anomaly θ_t at time t;
- Apply the transition matrices for in-plane and out-of-plane motions with (3.31) and (3.32);
- Revert coordinate transformations with (3.33) and (3.34) for $\theta = \theta_t$.

These transition matrices can be used as the dynamic matrix of a discrete linear time-variant system, setting the sampling period as $T_s = t - t_0$, and so they allow to easily simulate the relative motion between the target and chaser spacecraft in the LVLH frame for an elliptical orbit. This approach, however, only models free-drift motions, that is, in the absence of input forces. The particular solution for this problem in the domain of the true anomaly θ is presented in the next section.

3.3.4 Particular Solution

Rewriting the system of equations (3.27) in state form as $x' = Ax + Bu$, the particular solution is obtained with

$$x_p(\theta_t) = \int_{\theta_0}^{\theta_t} \Phi(\theta)B(\theta)u(\theta)d\theta , \tag{3.40}$$

where Φ represents the transition matrices presented in the previous section; for the in-plane motion $\Phi_i(\theta_0, \theta_t) = \phi_i(\theta_t)\phi_i^{-1}(\theta_0)$, and for the out-of-plane motion $\Phi_o(\theta_0, \theta_t) = \phi_o(\theta_0, \theta_t)$. Furthermore, for the in-plane motion, we have

$$B_i(\theta) = \frac{1}{m_c k^4 \rho^3(\theta)} \begin{bmatrix} 0 & 0 \\ 0 & 0 \\ 1 & 0 \\ 0 & 1 \end{bmatrix} \quad \text{and} \quad u_i(\theta) = \begin{bmatrix} F_x(\theta) \\ F_z(\theta) \end{bmatrix} , \tag{3.41}$$

and, for the out-of-plane motion,

$$B_o(\theta) = \frac{1}{m_c k^4 \rho^3(\theta)} \begin{bmatrix} 0 \\ 1 \end{bmatrix} \quad \text{and} \quad u_o(\theta) = F_y(\theta) . \tag{3.42}$$

The solution requires computing several non-trivial integrals, which has already been done in [14] for the case where the force is constant along the propagation interval ($\mathbf{F}(\theta) = \mathbf{F}$). The solution presented there for the in-plane motion is

$$\begin{bmatrix} \tilde{x}_p \\ \tilde{z}_p \\ \tilde{x}'_p \\ \tilde{z}'_p \end{bmatrix} = \Phi_i(\theta_0, \theta_t)\frac{1}{k^4(1-e^2)}\underbrace{\begin{bmatrix} 3I_{1J} - e(I_{s3} + I_{s2}) & 2I_3 - e(I_{c2} + 3I_{s2J}) \\ I_{s3} + I_{s2} - 3eI_{1J} & I_{c2} - e(2I_3 - 3eI_{s2J}) \\ I_{c3} + I_{c2} + eI_3 & -I_{s2} \\ -I_1 & eI_{s2} \end{bmatrix}}_{\Gamma_i(\theta_0, \theta_t)} \frac{1}{m_c}\begin{bmatrix} F_x \\ F_z \end{bmatrix} , \tag{3.43}$$

and for the out-of-plane motion is

$$\begin{bmatrix} \tilde{y}_p \\ \tilde{y}'_p \end{bmatrix} = \frac{1}{k^4}\underbrace{\begin{bmatrix} \sin(\theta_t) & \cos(\theta_t) \\ \cos(\theta_t) & -\sin(\theta_t) \end{bmatrix}\begin{bmatrix} I_{c3} \\ -I_{s3} \end{bmatrix}}_{\Gamma_o(\theta_0, \theta_t)} \frac{1}{m_c}F_y , \tag{3.44}$$

where the variables I_i are integrals, the values of which are presented in Appendix.

To obtain the full solution, the particular solution is added to the homogeneous solution. Merging the in-plane and out-of-plane motions yields the full solution,

$$
\begin{bmatrix} \tilde{x}_t \\ \tilde{y}_t \\ \tilde{z}_t \\ \tilde{x}'_t \\ \tilde{y}'_t \\ \tilde{z}'_t \end{bmatrix} = \Phi(\theta_0, \theta_t) \begin{bmatrix} \tilde{x}_0 \\ \tilde{y}_0 \\ \tilde{z}_0 \\ \tilde{x}'_0 \\ \tilde{y}'_0 \\ \tilde{z}'_0 \end{bmatrix} + \Gamma(\theta_0, \theta_t) \begin{bmatrix} F_x \\ F_y \\ F_z \end{bmatrix} (\theta_0),
\tag{3.45}
$$

where $\Phi(\cdot, \cdot)$ and $\Gamma(\cdot, \cdot)$ are appropriately generated from the entries of Φ_i, Φ_o, Γ_i and Γ_o. Inverting the coordinate transformation to obtain the solution in the time-domain yields

$$
\begin{bmatrix} x_t \\ y_t \\ z_t \\ \dot{x}_t \\ \dot{y}_t \\ \dot{z}_t \end{bmatrix} = \Lambda^{-1}(\theta_t)\Phi(\theta_0, \theta_t)\Lambda(\theta_0) \begin{bmatrix} x_0 \\ y_0 \\ z_0 \\ \dot{x}_0 \\ \dot{y}_0 \\ \dot{z}_0 \end{bmatrix} + \Lambda^{-1}(\theta_t)\Gamma(\theta_0, \theta_t) \begin{bmatrix} F_x \\ F_y \\ F_z \end{bmatrix}.
\tag{3.46}
$$

Note that, since for the particular solution, a constant force between sampling intervals is assumed, this solution constitutes a zero-order hold (ZOH) input discretization.

3.3.4.1 State-Space Model

Equation (3.46) can be used as a discrete state model of a linear and time-variant system

$$
\mathbf{x}_{k+1} = A_k^{k+1}\mathbf{x}_k + B_k^{k+1}\mathbf{u}_k,
\tag{3.47}
$$

where the state vector is $\mathbf{x} = [x, y, z, \dot{x}, \dot{y}, \dot{z}]^\top$ and the input vector $\mathbf{u} = [F_x, F_y, F_z]^\top$, and such that the system at time k has true anomaly θ_0 and at time $k+1$ the true anomaly θ_t. Matrix A_k^{k+1} is the state transition matrix from time k to $k+1$, and from (3.46) is defined as

$$
A_k^{k+1} = \Lambda^{-1}(\theta_t)\Phi(\theta_0, \theta_t)\Lambda(\theta_0),
\tag{3.48}
$$

while B_k^{k+1} is the input matrix, which becomes

$$
B_k^{k+1} = \Lambda^{-1}(\theta_t)\Gamma(\theta_0, \theta_t).
\tag{3.49}
$$

3.4 Relative Motion in a Circular Target Orbit

In this section and the next, the relative motion between two satellites is simulated, considering circular and elliptical target orbits, respectively. Both free-drift motions and impulsive thrust manoeuvres are presented.

The relative motion for a circular target orbit can be simulated with the Hill linearized dynamics and can either be numerically simulated in Simulink, with the continuous state-space model in (3.13) and (3.14) or analytically simulated with its solution, the Clohessy–Wiltshire equations. Because numerical simulation introduces a discretization error, it is preferable to use the analytical solution. All simulations are performed for Earth satellites with a target orbit height of 600 km from the surface, simulated for two orbital periods and with a sampling period of 10 s. The orbit orientation parameters are disregarded, since a uniform gravitational field is assumed, and the Hill equations describe the relative motion on the orbital plane.

Despite all satellite trajectories being circles or ellipses on the inertial frame, the relative trajectories on this frame may be non-intuitive, since they are non-inertial. On the R-bar/V-bar plots, the target spacecraft is always on the origin; hence, if the chaser travels along the negative direction of R-bar (up), then it gains altitude relative to the target, and if it travels along the positive direction of V-bar (left), then it gets ahead of the target in its orbit. Because the in-plane and out-of-plane motions are decoupled in the linearised model, they will be shown separately. The chaser initial position is marked with '×', while the target position (origin) is marked with 'O'.

3.4.1 Free-Drift Motions

To begin with, the motion with no action from the chaser actuators is considered, simulated with different initial conditions for position and velocity.

If the chaser lies on V-bar (R-bar $= 0$) with the same velocity as the target, it will not move relative to it, as shown in Fig. 3.3, since in these conditions the spacecraft are simply on a different phase of *approximately* the same circular orbit.

If the chaser is at a different altitude and with zero relative velocity, then it is necessarily on an elliptic orbit, since higher/lower altitude circular orbits have lower/higher orbital velocity. Hence, as seen in Fig. 3.4, if the target starts at a higher altitude, it will gain altitude until it reaches apogee, and then comes back to perigee. On the other hand, if it starts lower, it will lose altitude and gain velocity until it reaches perigee, and then climbs back up to apogee. Moreover, the chaser at a higher orbit will have a greater orbital period, and so will fall behind the target, while the one with the lower orbit will get ahead.

If the horizontal velocity is compensated as to ensure the chaser is on a circular orbit, it will now just drift along a constant altitude, as shown in Fig. 3.5. For a difference in orbit heights of z_0, the chaser relative horizontal velocity that generates a circular chaser orbit is $V_x = \frac{3}{2}\omega z_0$.

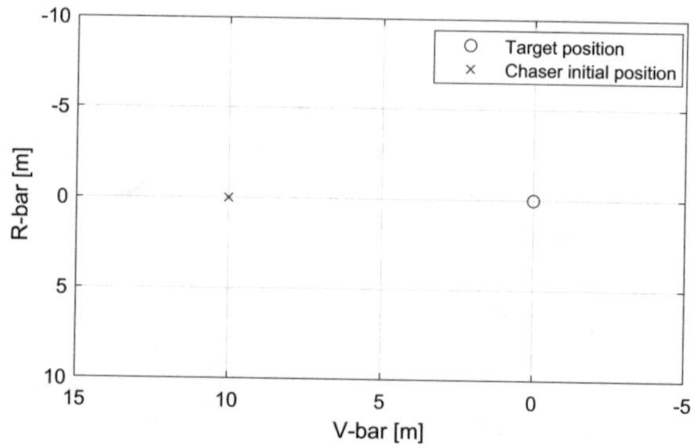

Fig. 3.3 Relative in-plane motion on a circular target orbit with V-bar start. Initial conditions $s_0 = [10, 0, 0]$ m, $\dot{s}_0 = [0, 0, 0]$ m/s

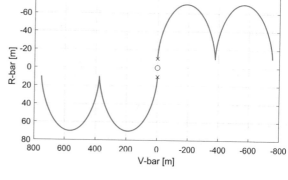

Fig. 3.4 Relative in-plane motion on a circular target orbit with R-bar start. Initial conditions $s_0 = [0, 0, \pm 10]$ m, $\dot{s}_0 = [0, 0, 0]$ m/s

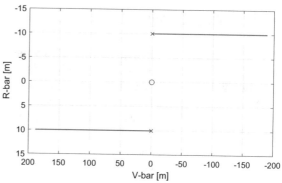

Fig. 3.5 Relative in-plane motion in a circular target orbit with R-bar start and circular chaser orbit. Initial conditions $s_0 = [0, 0, \pm 10]$ m, $\dot{s}_0 = [\pm 10 \frac{3}{2}\omega, 0, 0]$ m/s

In Fig. 3.6, the chaser starts at V-bar with a lower radial velocity, which causes it to be on an elliptic orbit, however, with the same orbital period as the target. This causes the chaser to drop down and get ahead of the target and then looping around to the same initial relative position.

If the chaser starts on V-bar with a higher horizontal velocity, it will get ahead of the target and gain altitude. As it does, the chaser loses velocity and starts falling behind, until it reaches apogee. It will then lose altitude and gain velocity, and then

Fig. 3.6 Relative in-plane motion in a circular target orbit with V-bar start and radial relative velocity. Initial conditions
$s_0 = [-10, 0, 0]$ m,
$\dot{s}_0 = [0, 0, 0.01]$ m/s

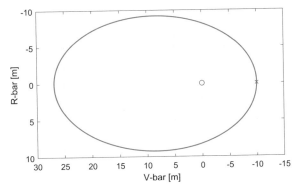

Fig. 3.7 Relative in-plane motion in a circular target orbit with V-bar start and horizontal relative velocity. Initial conditions
$s_0 = [\pm 10, 0, 0]$ m,
$\dot{s}_0 = [\pm 0.01, 0, 0]$ m/s

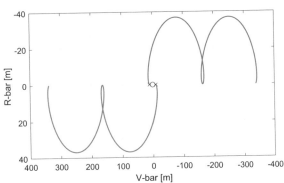

start catching up again as it reaches perigee. This causes the loops seen in Fig. 3.7 around perigee. If the chaser starts with a lower horizontal velocity, the trajectory will be mirrored.

The out-of-plane motion without any actuation will always be sinusoidal with respect to time. In Fig. 3.8, the chaser starts above the target orbital plane (H-bar=0), with the same normal velocity. It will then decrease along H-bar and cross the target orbital plane at the ascending node until it reaches the opposite distance with which it started. It will then increase and intersect the orbital plane at the descending node until it reaches the initial position.

3.4.2 Impulsive Thrust Manoeuvres

We will now consider manoeuvres in which instant changes in velocity (ΔV) are applied by the chaser actuators along the trajectory. Note that, in reality, these instant ΔV's are impossible, since any spacecraft thrusters can only generate a gradual increase in velocity. Nevertheless, it is useful to consider this type of manoeuvres for

Fig. 3.8 Relative
out-of-plane motion in a
circular target orbit. Initial
conditions $s_0 = [0, 10, 0]$ m,
$\dot{s}_0 = [0, 0, 0]$ m/s

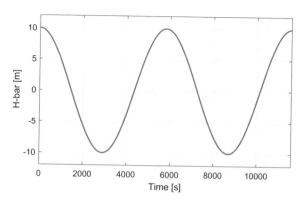

Fig. 3.9 Hohmann transfer
manoeuvre in a circular
target orbit

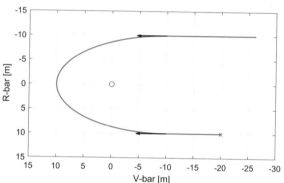

analysis and planning of rendezvous trajectories. The deduction of the expressions
for the ΔV's can be found in the book by Fehse [10].

In Fig. 3.9, a Hohmann transfer manoeuvre is shown relative to the target, in which
two ΔV's (presented as black arrows) are applied in order to change the altitude of
the chaser by Δz. The chaser starts on a circular orbit below the target and then
increases its horizontal velocity by

$$\Delta V_x = \Delta z \omega / 4, \tag{3.50}$$

which changes its orbit to an eccentric transfer trajectory with the apogee at the
desired altitude. At apogee, the same ΔV is applied in order to circularize the orbit.
Thus, it takes half an orbital period to complete this manoeuvre.

One way to perform a V-bar transfer, in which the chaser advances or retreats by
Δx, is with two radial ΔV's with magnitude

$$\Delta V_z = \Delta x \omega / 4, \tag{3.51}$$

exploiting a trajectory such as the one seen in Fig. 3.6. In Fig. 3.10, the chaser is on
the same orbit but behind the target, which causes it to move below and ahead of

Fig. 3.10 Radial V-bar transfer manoeuvre with radial impulses in a circular target orbit

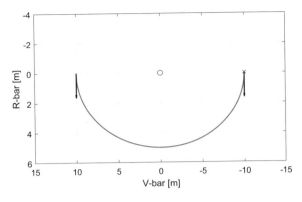

Fig. 3.11 Radial V-bar transfer manoeuvre with horizontal impulses in a circular target orbit

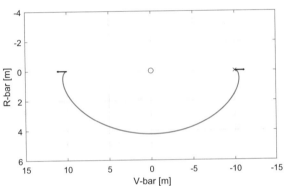

the target. As it crosses V-bar, it executes the same ΔV to circularize the orbit and remain stationary relative to the target.

Another possible V-bar transfer manoeuvre is with the use of horizontal ΔV's (instead of radial) with magnitude

$$\Delta V_x = -\Delta x \frac{\omega}{6\pi}, \tag{3.52}$$

exploiting a trajectory like the one in Fig. 3.7, and resulting in the manoeuvre in Fig. 3.11. This manoeuvre takes one full orbital period to complete, as opposed to half a period with the radial impulses, but costs $3\pi/2$ times less, which is very significant.

To fully correct a chaser inclination difference in respect to the target orbit requires a normal ΔV at either the ascending or descending node. If the amplitude of the out-of-plane motion is Δy, then the required ΔV is

$$\Delta V_y = -\omega \Delta y. \tag{3.53}$$

Figure 3.12 shows an example of this manoeuvre.

Fig. 3.12 Inclination
correction manoeuvre in a
circular target orbit

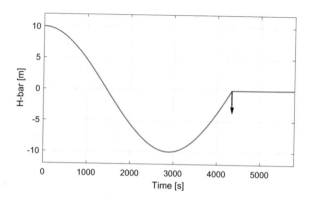

3.5 Relative Motion in an Elliptic Target Orbit

The relative motion in an elliptic target orbit was simulated using the discrete state-space model in (3.47). Because the orbit is elliptic, the initial true anomaly θ_0 now has to be defined as an initial condition. As before, the simulations are performed for an Earth satellite, with a perigee height of 600 km from the surface, and simulated for two orbital periods and a sampling true anomaly Θ_s of 0.5°. The eccentricity of the orbit will vary between experiments. As before, the orbit orientation parameters are disregarded.

The relative motion in an elliptic orbit is significantly more complex and harder to comprehend. One reason behind this is that for elliptic orbits the LVLH frame rate of rotation varies along the orbit, while for circular orbits, it rotates uniformly. Also note that, for elliptic orbits, the V-bar axis is not always aligned with the target velocity vector, unlike for circular orbits.

3.5.1 Free-Drift Motions

As before, the motions in the absence of thrust will first be addressed. With a V-bar start and zero initial relative velocity, it is shown in Fig. 3.13 that the chaser drifts away from the target, unlike what happens for the same conditions in a circular orbit, where it would remain stationary (Fig. 3.3). This happens because the magnitude of the orbital velocity is not constant along an elliptic orbit, which means that if the chaser is ahead and with the same velocity, then it is not exactly on the same orbit as the target, and thus the spacecraft present some relative motion. It is also remarked that, since V-bar is not necessarily aligned with the target velocity vector, being on the same orbit may require the chaser to be on a different R-bar position. It may also be observed that the drift trajectory changes with the position of the target along its orbit, since the dynamics change with the true anomaly θ; if the target starts at perigee ($\theta_0 = 0°$) or apogee ($\theta_0 = 180°$), the drift happens only along V-bar, otherwise the

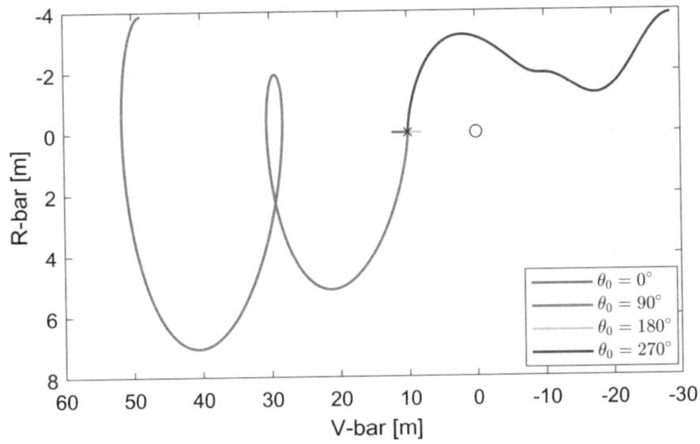

Fig. 3.13 Relative in-plane motion in an elliptic target orbit with V-bar start and various initial true anomalies. Initial conditions $s_0 = [10, 0, 0]$ m and $\dot{s}_0 = [0, 0, 0]$ ms^{-1}, with eccentricity $e = 0.1$

Fig. 3.14 Relative in-plane motion in an elliptic target orbit with V-bar start and different initial true anomalies. Initial conditions $s_0 = [10, 0, 0]$ m and $\dot{s}_0 = [0, 0, 0]$ ms^{-1}, with eccentricity $e = 0.5$

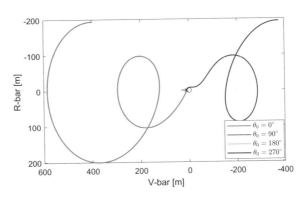

chaser also drifts in R-bar. Increasing the eccentricity, the drift motions now have greater amplitudes, as shown in Fig. 3.14.

If the initial relative velocity is compensated to ensure the spacecraft are on the same orbit, but with different true anomalies, the results from Fig. 3.15 are obtained, where two different eccentricities are simulated. To generate these initial conditions, the chaser and target positions were defined on the orbital plane frame with the orbit elements, on the same orbit but with different phases, and then transformed to the target local orbital frame with the transformation (3.5), as well as (3.6) for the velocity. Despite being on the same orbit, the spacecraft still move relative to each other, although it can be observed that, due to the fact that the spacecraft are on the same orbit and thus have the same orbital period, the chaser returns to the initial relative position after one orbit. It can be again observed that a higher eccentricity leads to a greater amplitude of the relative motion.

The motion along R-bar is due to the fact that, as mentioned before, the V-bar axis is not always aligned with the velocity vector, as can be observed in Fig. 3.2, and so it moves relative to it along the orbit: at perigee, they are aligned, and then V-bar lowers toward the inside of the ellipse until it reaches perigee and they become aligned again, after which V-bar raises toward the outside of the ellipse. Since in Fig. 3.15 the target starts at perigee, the chaser first decreases in R-bar, and then it comes back up until

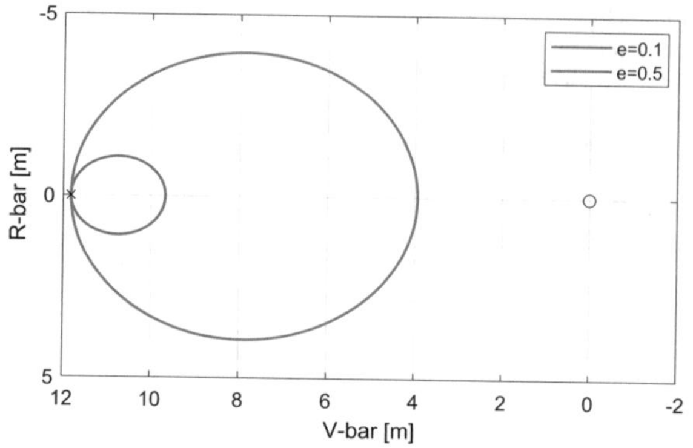

Fig. 3.15 Relative in-plane motion in an elliptic target orbit with chaser on target orbit. Initial true anomaly $\theta_0 = 0°$ and initial orbital phase of $\Delta\theta = 0.0001°$

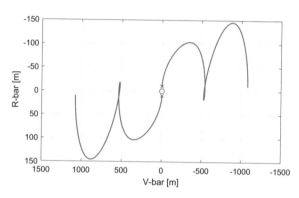

Fig. 3.16 Relative in-plane motion in an elliptic target orbit with R-bar start. Initial conditions $s_0 = [0, 0, \pm 10]$ m, $\dot{s}_0 = [0, 0, 0]$ ms, $\theta_0 = 0°$, eccentricity $e = 0.1$

it reaches perigee and increases R-bar. On the other hand, the motion along V-bar is due to the varying orbital velocity. As the spacecraft move toward apogee they lose velocity, and, because the chaser is ahead, the target catches up with it and closes their relative distance. When moving toward perigee, the spacecraft speed up and the chaser gets ahead of the target, increasing along V-bar.

With an R-bar perigee start with zero relative velocity, the result in Fig. 3.16 is obtained. The drift motion is similar to that of a circular target orbit in the same conditions (Fig. 3.4), but now the amplitude of the trajectory increases with each orbit. If the chaser starts at $\theta_0 = 90°$ instead, the motion is different, as shown in Fig. 3.17.

If the chaser starts on a value of R-bar on a higher or lower orbit, but such that it has the same eccentricity as the target orbit, the result in Fig. 3.18 is obtained. Unlike in the case of a circular orbit with the same initial conditions (Fig. 3.5), where the chaser simply drifts along V-bar, now it also moves along R-bar. This is in part due to the motion of V-bar relative to the target velocity vector but also because two orbits with the same eccentricity and a different perigee height will not have the same difference in apogee heights. Also, because the satellites have different orbital periods, the amplitude of the motion increases as they phase along the orbit. The initial conditions for this simulation were again defined in the orbital plane

Fig. 3.17 Relative in-plane motion in an elliptic target orbit with R-bar start and different initial true anomaly. Initial conditions $s_0 = [0, 0, \pm 10]$ m, $\dot{s}_0 = [0, 0, 0]$ ms^{-1}, $\theta_0 = 90°$, eccentricity $e = 0.1$

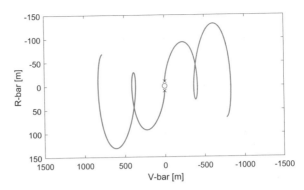

Fig. 3.18 Relative in-plane motion in an elliptic target orbit with chaser on lower/higher with target eccentricity. Initial true anomaly $\theta_0 = 0°$, eccentricity $e = 0.1$ and difference in semi-major axis of $\Delta a = \pm 11$ m

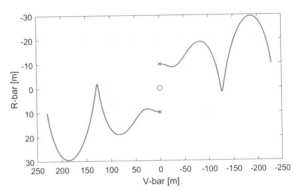

frame, placing the chaser on an orbit with the same eccentricity but with a different semi-major axis and on the same true anomaly, and then converting to the LVLH frame.

For the out-of-plane motion, the chaser trajectory is still approximately a sine wave, although the resemblance fades with the increase of the eccentricity, as can be observed in Fig. 3.19. It can also be concluded from the results that the trajectory depends on the initial position of the target along its orbit, i.e. with the initial θ.

At perigee ($\theta = 0°$), the acceleration in H-bar is greater than at apogee ($\theta = 180°$), and so if the spacecraft start at perigee (blue trajectory) the chaser will drop further down along the normal direction than if they start at apogee (yellow trajectory). Furthermore, because of the difference in H-bar acceleration, the spacecraft will spend more time at apogee than at perigee, which causes the peaks of the blue trajectory to be more narrow than the valleys, while the opposite happens for the yellow trajectory. This effect is greater for a bigger eccentricity. If the chaser does not start at perigee or apogee (orange trajectory), the peaks and valleys do not match with apogee and perigee, and so the varying dynamics cause the wave to be asymmetrical.

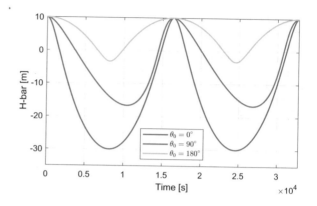

Fig. 3.19 Relative out-of-plane motion in an elliptic target orbit with increased eccentricity. Initial conditions $s_0 = [0, 10, 0]$ m, $\dot{s}_0 = [0, 0, 0]$ ms^{-1} and different initial true anomalies, eccentricity $e = 0.5$

3.5.2 Impulsive Thrust Manoeuvres

The ΔV for an arbitrary impulsive manoeuvre given specific initial and final conditions can be determined with the Yamanaka–Ankersen transition matrix. Using the coordinate transformations we have for the in-plane motion

$$\begin{bmatrix} x_f \\ z_f \\ \dot{x}_f \\ \dot{z}_f \end{bmatrix} = \Lambda_i^{-1}(\theta_f)\Phi_i(\theta_f)\Lambda_i(\theta_0) \begin{bmatrix} x_0 \\ z_0 \\ \dot{x}_0 \\ \dot{z}_0 \end{bmatrix}, \qquad (3.54)$$

where the index f represents the final conditions, and the transition time between the initial and final positions is specified with θ_f. Given an initial position, the required initial velocity to achieve the final position at the specified time can be obtained by solving the above equation for $\dot{x}_0$ and $\dot{z}_0$. Denoting the transformed transition matrix as D, the first two equations yield

$$\begin{bmatrix} x_f \\ z_f \end{bmatrix} = \begin{bmatrix} d_{11} & d_{12} & d_{13} & d_{14} \\ d_{21} & d_{22} & d_{23} & d_{24} \end{bmatrix} \begin{bmatrix} x_0 \\ z_0 \\ \dot{x}_0 \\ \dot{z}_0 \end{bmatrix}, \qquad (3.55)$$

and so the required ΔVs are

$$\dot{x}_0 = \frac{d_{14}z_f + (d_{24}d_{11} - d_{21}d_{14})x_0 + (d_{24}d_{12} - d_{22}d_{14})z_0 - d_{24}x_f}{d_{23}d_{14} - d_{24}d_{13}},$$

$$\dot{z}_0 = \frac{d_{23}x_f + (d_{13}d_{21} - d_{11}d_{23})x_0 + (d_{13}d_{22} - d_{12}d_{23})z_0 - d_{13}z_f}{d_{23}d_{14} - d_{24}d_{13}}. \qquad (3.56)$$

Note that, due to the denominator in the above expressions, a transition time equal to an orbital period generates a singularity, and so is not possible. To eliminate

Fig. 3.20 R-bar transfer manoeuvre in an elliptic target orbit. Initial position $s_0 = [-10, 0, 10]$ m at $\theta_0 = 0°$, final position $s_f = [-10, 0, -10]$ m at $\theta_f = 180°$, with eccentricity $e = 0.4$

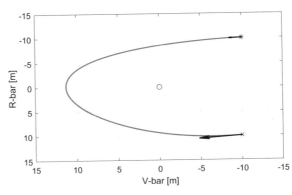

Fig. 3.21 Arbitrary in-plane transfer manoeuvre in an elliptic target orbit. Initial position $s_0 = [-75, 0, -15]$ m at $\theta_0 = 0°$, final position $s_f = [10, 0, -40]$ m at $\theta_f = 180°$, with eccentricity $e = 0.4$

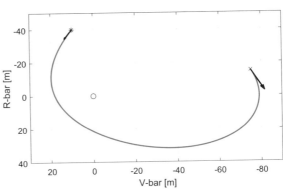

the relative velocity at the end of the transfer, the final ΔV can be determined by computing the negative of $\dot{x}_f$ and $\dot{z}_f$ from Eq. (3.54).

Figure 3.20 shows an example of a V-bar transfer manoeuvre, similar to the one seen in Fig. 3.9 for a circular target orbit. It can be observed that the chaser achieves the final position in the specified transfer time. Figure 3.21 shows an in-plane transfer between arbitrary points.

For the out-of-plane dynamics, Eq. (3.54) becomes

$$\begin{bmatrix} y_f \\ \dot{y}_f \end{bmatrix} = \Lambda_o^{-1}(\theta_f)\Phi_o(\theta_f)\Lambda_o(\theta_0)\begin{bmatrix} y_0 \\ \dot{y}_0 \end{bmatrix}, \qquad (3.57)$$

which yields the ΔV

$$\dot{y}_0 = \frac{k^2 \rho(\theta_0)}{\sin(\theta_f - \theta_0)}\left[\rho(\theta_f)y_f - (\cos(\theta_f - \theta_0) + e\cos(\theta_f))y_0\right]. \qquad (3.58)$$

This expression is singular for $\theta_f - \theta_0 = n\pi$, $n \in \mathbb{N}$, and thus transfers of half or full orbit are not possible.

Notice how for all these manoeuvres the first ΔV is greater than the last. This is due to the fact that these manoeuvres end at/near apogee, where the dynamics are slower than at perigee.

Appendix: Particular Solution Integrals

In this appendix, the expressions for the integrals in Eqs. (3.43) and (3.44) are presented, as solved in [14].

$$I_{s_3} = \frac{1}{2e}\left[\frac{1}{\rho(\theta_t)^2} - \frac{1}{\rho(\theta_0)^2}\right]. \tag{3.59}$$

$$I_{c_3} = (1-e^2)^{-\frac{5}{2}}\left[(1+e^2)(\sin(E_t) - \sin(E_0)) - \frac{e}{2}(\sin(E_t)\cos(E_t) - \right.$$
$$\left. - \sin(E_0)\cos(E_0) + 3(E_t - E_0))\right], \tag{3.60}$$

$$I_{s_2} = \frac{1}{e}\left[\frac{1}{\rho(\theta_t)} - \frac{1}{\rho(\theta_0)}\right], \tag{3.61}$$

$$I_{c_2} = (1-e^2)^{-\frac{3}{2}}\left[\sin(E_t) - \sin(E_0) - e(E_t - E_0)\right], \tag{3.62}$$

$$I_3 = (1-e^2)^{-\frac{5}{2}}\left[(\frac{1}{2}e^2 + 1)(E_t - E_0) + \right.$$
$$\left. + \frac{1}{2}e^2(\sin(E_t)\cos(E_t) - \sin(E_0)\cos(E_0)) - 2e(\sin(E_t) - \sin(E_0))\right], \tag{3.63}$$

$$I_1 = (1-e^2)^{-\frac{1}{2}}(E_t - E_0), \tag{3.64}$$

$$I_{1J} = (1-e^2)^{-2}\left[\frac{1}{2}(E_t^2 - E_0^2) + e(\cos(E_t) - \cos(E_0)) + (e\sin(E_0) - E_0)(E_t - E_0)\right], \tag{3.65}$$

$$I_{s_{2J}} = (1-e^2)^{-\frac{5}{2}}\left[\sin(E_t)(1 + \frac{e}{2}\cos(E_t)) - E_t(\frac{e}{2} + \cos(E_t)) - \right.$$
$$\left. \sin(E_0)(1 + \frac{e}{2}\cos(E_0)) + E_0(\frac{e}{2} + \cos(E_0)) - (e\sin(E_0) - E_0)(\cos(E_t) - \cos(E_0))\right]. \tag{3.66}$$

References

1. G.W. Hill, Researches in the lunar theory. Am. J. Math. **1**, 5–26 (1878). ISSN: 00029327, 10806377
2. K. Yamanaka, F. Ankersen, New state transition matrix for relative motion on an arbitrary elliptical orbit. J. Guid. Control Dyn. **25**, 60–66 (2002)
3. C. Wei, S.-Y. Park, C. Park, Linearized dynamics model for relative motion under a J2-perturbed elliptical reference orbit. Int. J. Non-Linear Mech. **55**, 55–69 (2013)
4. L. Cao, A.K. Misra, Linearized J2 and atmospheric drag model for satellite relative motion with small eccentricity. Proc. Inst. Mech. Eng. Part G: J. Aerosp. Eng. **229**, 2718–2736 (2015)
5. L. Breger, J.P. How, Gauss's variational equation-based dynamics and control for formation flying spacecraft. J. Guid. Control Dyn. **30**, 437–448 (2007)
6. S. D'Amico, Relative orbital elements as integration constants of Hill's equations. DLR, TN, 05-08 (2005)
7. A.W. Koenig, T. Guffanti, S. D'Amico, New state transition matrices for spacecraft relative motion in perturbed orbits. J. Guid. Control Dyn. **40**, 1749–1768 (2017)
8. K. Alfriend, H. Yan, Evaluation and comparison of relative motion theories. J. Guid. Control Dyn. **28**, 254–261 (2005)
9. J. Sullivan, S. Grimberg, S. D'Amico, Comprehensive survey and assessment of spacecraft relative motion dynamics models. J. Guid. Control Dyn. **40**, 1837–1859 (2017)
10. W. Fehse, *Automated Rendezvous and Docking of Spacecraft* (Cambridge University Press, 2003). ISBN: 0521824923
11. W.H. Clohessy, R.S. Wiltshire, Terminal guidance system for satellite rendezvous. J. Aerosp. Sci. **27**, 653–658 (1960)
12. J. Tschauner, P. Hempel, Rendezvous with a target in an elliptical orbit. Astronaut. Acta **11**, 104–109 (1965)
13. D.F. Lawden, Fundamentals of space navigation. J. Br. Interplanet. Soc. **13**, 87–101 (1954)
14. F. Ankersen, Guidance, navigation, control and relative dynamics for spacecraft proximity maneuvers. Ph.D. thesis, Institut for Elektroniske Systemer (2010). ISBN: 9788792328724

Chapter 4
Rendezvous with Model Predictive Control

In traditional rendezvous mission design, guidance trajectories are designed offline, and so manoeuvres are performed in open-loop, often with punctual mid-course correction boosts determined online from the trajectory deviation [1]. The advantage of this approach lies in the low complexity and computational load required on board, which is desirable due to the typically limited computational resources available. In this context, with the improved computational capability available came an increasing amount of research dedicated to applying Model Predictive Control (MPC) to the rendezvous guidance problem [2–10], in order to perform these thrust manoeuvres online and in full closed loop. This feature is desirable since it increases the autonomy of the spacecraft and allows for more precise and fuel-efficient manoeuvres. Furthermore, MPC can handle crucial operational constraints present in a rendezvous mission, such as

- minimization of propellant consumption,
- limited thruster authority,
- spacecraft collision safety and passive safety.

Most of the MPC literature for rendezvous is dedicated solely to the translational control of the spacecraft. In fact, in rendezvous processes, the attitude and position control are typically treated separately since there is a weak coupling between translational and rotational motion [1]. Nevertheless, docking and berthing operations require the two controllers to be coupled, although these manoeuvres are outside the scope of this work. Furthermore, the spacecraft attitude may be subject to more operational constraints, such as the exposure of the solar panels to the sun, or the orientation of antennae toward ground stations, that significantly complicate the MPC optimization problem. Although MPC has been applied to spacecraft attitude control [11], and to coupled control [12–16], this book only deals with translational control.

MPC can perform both the guidance and control functions of a spacecraft GNC system. The manoeuvre terminal state can be input as the controller reference, generating a trajectory much like a guidance system. It also generates a sequence of control decisions, that can be discarded in favour of a different low-level and higher frequency

© The Author(s), under exclusive license to Springer Nature Switzerland AG 2021
A. Botelho et al., *Predictive Control for Spacecraft Rendezvous*,
SpringerBriefs in Applied Sciences and Technology,
https://doi.org/10.1007/978-3-030-75696-3_4

controller, or used as a feed-forward control action for that controller. MPC can also perform solely controller functions if a reference trajectory is provided, rather than a reference terminal state, although that is not explored here. Nevertheless, the MPC algorithms designed in this book will still be referred to as a 'controller', despite being akin to a traditional spacecraft guidance system, and this does not invalidate the need for a low-level and higher frequency controller and attitude controller. The navigation function is out of scope for this work, and the state is always assumed to be known, although Sect. 4.6 does consider the presence of state estimation errors.

The computational requirement for MPC is its greatest limitation. In a rendezvous context, however, it can be computationally feasible to implement MPC in real-time, due to the fact that orbital dynamics are considerably slow, and that relative dynamics can be accurately linearized, as shown in Chap. 3, that allows for the use of Linear MPC. The MPC computation time will be a major consideration in this chapter. Furthermore, since it is proposed to function as a guidance algorithm, the MPC algorithm presented will run at a low frequency, which further improves its real-time computational feasibility.

In Sect. 4.1, we present a method for sampling the dynamics for the prediction model, which deals with the fact that orbital dynamics are highly time-varying for highly elliptical orbits. Section 4.2 applies the standard MPC approach with the receding horizon strategy to the rendezvous problem, showing how it is not appropriate for use in this application. Instead, Sect. 4.3, presents the *finite-horizon* strategy and shows it to be more appropriate for rendezvous since it allows for fuel-optimal manoeuvres. Section 4.4 presents the alternative *variable-horizon* MPC formulation, which allows for optimizing manoeuvre duration simultaneously with fuel. In Sect. 4.5 we approach the passive safety problem, presenting two new techniques for an efficient implementation of these constraints for real-time optimization. Section 4.6 considers the presence of perturbations and disturbances and presents several robustness techniques, of which some are first proposed in this work. Finally, Sect. 4.7 presents several simulations with the methods presented in this Chapter. As a case study for scenarios with highly elliptical orbits, we will consider the conditions of the future PROBA-3 rendezvous experiment ESA mission [17], and simulations will be performed in a high-fidelity industrial simulator.

4.1   Prediction Model and Relative Dynamics Sampling

The prediction model to be used in the MPC formulation is the Yamanaka–Ankersen state transition matrix [18] together with the Ankersen ZOH particular solution [19], presented and derived in Chap. 3, that provides a linear model of the relative dynamics between two spacecraft in an elliptic orbit and with Cartesian coordinates. The use of a model with ZOH control discretization, as opposed to the typically utilized impulsive ΔV models, allows the guidance to generate more realistic reference trajectories and control profiles, that take into account the length of time it takes to apply a control command. On the other hand, for impulsive ΔV guidance planners, the solution has to be discretized into an a posteriori force-profile, which can be a source of error.

It is remarked that, as previously mentioned in Chaps. 1 and 3, alternative linear dynamic models are available [20, 21]. These include models with perturbations such as J_2 and atmospheric drag [22, 23]. Other noteworthy examples include models based on Relative Orbital Elements (ROEs) [24–27], rather than Cartesian coordinates, which can readily include certain environmental perturbations, such as J_2, and which remain accurate for larger relative distances [2]. An alternative ROE formulation has been presented by D'Amico et al. [28–30], that provides geometric insight into the relative trajectory given a near-circular target orbit, which is especially useful for designing passively safe configurations. More recently these have been extended to also include atmospheric drag as well as J_2 [31]. Finally, models that consider restricted three-body dynamics, namely near-rectilinear halo orbits (NRHOs), are also available [32].

Nevertheless, one feature of MPC is that, given that the optimal control problem is numerically computed in real time, the methodology is incredibly versatile with respect to the model and the selected prediction model may be easily swapped with another by simply changing the state model matrices. Therefore, nearly everything discussed and covered in this book is still applicable to most other linear relative orbital dynamic models available, and choice of the Yamanaka–Ankersen equations for the prediction model presents no loss of generality. One exception is the collision safety and passive safety constraints, which are not directly applicable to the case in which different state variables, such as ROEs, are employed.

Regardless of the model selected, a difficulty arises related to the fact that orbital dynamics are time-varying for an elliptic orbit. For example, in the conditions of the ESA PROBA-3 mission [17], where the perigee height is 600 km and the orbit is highly elliptical with an eccentricity of 0.8111, the orbital period is approximately 19.6 h. Despite this very long period, in just 100 s the spacecraft will change their true anomaly by 8.3° from perigee, while in the same amount, it only changes 0.091° from apogee, as illustrated in Fig. 4.1. Thus, in these conditions, the dynamics are about 100 times faster at perigee than at apogee, which then translates to the velocity of the relative motion.

Because for MPC there is a limited amount of samples available, associated with the length of the prediction horizon, these must be allocated appropriately along the orbit in order to get the best performance since the point at which the thrust is applied is important for the optimality of the trajectory. If the dynamics are sampled with constant time intervals, as is standard, with 100 samples and in the conditions of the PROBA-3 mission, we get the result in Fig. 4.2a. The samples concentrate on apogee because the orbital velocity is lower there, which is opposite to what is desired, since the dynamics are faster on perigee. This results in the true anomaly interval between samples being over 50° at perigee, and less than 1° at apogee.

An alternative is to sample the dynamics with a constant true anomaly. This has the opposite effect, as seen in Fig. 4.2b, where now the samples are concentrated on perigee, which is more desirable. One further but unintentional advantage is that Kepler's equation no longer needs to be solved in order to determine the true anomaly

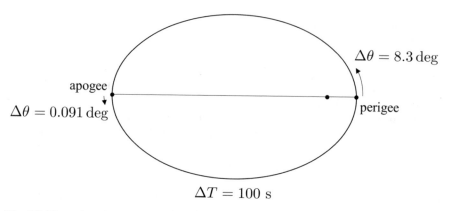

Fig. 4.1 Illustration of the time variation of the relative dynamics, in the conditions of the PROBA-3 mission

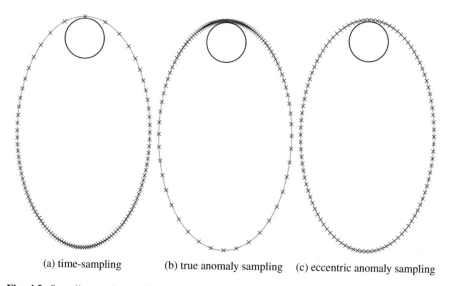

(a) time-sampling   (b) true anomaly sampling   (c) eccentric anomaly sampling

Fig. 4.2 Sampling methods of relative dynamics with 100 samples in conditions of PROBA-3 mission

and compute the transition matrix. However, the samples at apogee may now be too spread apart, and the time interval is almost 4000 s at apogee, and 40 s at perigee.

A final possibility is to sample the system with constant eccentric anomaly intervals. As seen in Fig. 4.2c, this procedure results in the samples being evenly spread in space. Furthermore, the time intervals are still greater at apogee than at perigee, but the contrast is not as considerable as before, with 1250 s at apogee and 170 s at perigee. This approach might thus be the most appropriate and will be used throughout this chapter. Note that sampling the system without time-constant intervals has an effect on the actuation profile, since a ZOH discretization is utilized, which results in a constant thrust along with each sampling interval, implying that each thrust action will have a different duration. In the case of a circular target orbit, the dynamics are time-invariant, and so time sampling will be used. Since manoeuvres can last up to several hours, sampling periods are often very large, and thus the ZOH discretization results in very long constant burns being commanded, which may be undesirable. However, the burn time in the particular solution in (3.40) can be set to a value smaller than the sampling period, yielding a partial zero-order hold discretization, although this option is not followed in this work.

4.2 Rendezvous with Receding-Horizon Control

The most common and naive approach for the MPC formulation is to use a quadratic cost function, as presented in Eq. (2.12), together with the standard receding horizon strategy. In the absence of state and control constraints, it becomes

$$
\min_{\substack{\bar{u}_0,\dots,\bar{u}_{N-1} \\ \bar{x}_0,\dots,\bar{x}_N}} \quad \sum_{i=0}^{N-1} (\bar{x}_i - x_{ref})^\top Q(\bar{x}_i - x_{ref}) + \bar{u}_i^\top R\bar{u}_i + \tag{4.1a}
$$
$$
+ (\bar{x}_N - x_{ref})^\top Q_f(\bar{x}_N - x_{ref}),
$$
$$
\text{s.t.} \quad \bar{x}_0 = x_t, \tag{4.1b}
$$
$$
\bar{x}_{k+1} = A_k^{k+1}\bar{x}_k + B_k^{k+1}\bar{u}_k, \ k = 0,\dots,N-1, \tag{4.1c}
$$

where the prediction model is that presented in (3.47).

This section will show through simulation that the receding horizon strategy without further changes is not appropriate for the rendezvous problem, as well as show the progression toward an optimal formulation. To simplify, a circular target orbit will be considered, with the height of the perigee of the PROBA-3 mission (600 km), and the chaser mass considered is that of the PROBA-3 Occulter Spacecraft, with 211 kg (launch mass). A simple V-bar transfer manoeuvre of 30 m is considered, in order to evaluate the controller performance by comparing it with the ideal impulsive manoeuvres of this type, shown in Figs. 3.10 and 3.11. The controller parameters used in the following simulations, as well as the ΔV spent performing the manoeuvre and the average computation time, are presented in Table 4.1. In the absence of any state or control constraints, the QP can be solved analytically.

Also, because the in-plane and out-of-plane motions are decoupled, the two can be solved in separate MPC problems, where each problem becomes smaller and easier to solve. Although solving separately was found to improve the computation time when solving analytically, it often became worse when solving numerically. Furthermore, this separation is not possible in the presence of constraints that relate both motions, such as collision avoidance and passive safety constraints.

Table 4.1 Controller parameters and results for V-bar transfer manoeuvre simulations with receding-horizon quadratic MPC

Figures	T_s (s)	N	R	Q	Q_f	ΔV	t_{avg} (μs)
4.3	1	10	I	$100I$	$100I$	6.77 m/s	40
4.4	1	10	I	I	I	1.41 m/s	39
4.5	100	10	$10^6 I$	I	I	111 mm/s	50
4.6	290	20	I	0	$100I$	5.27 mm/s	70

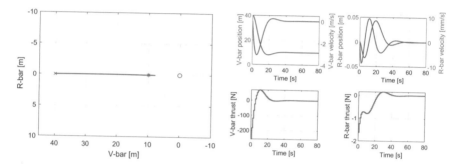

Fig. 4.3 V-bar transfer manoeuvre with receding-horizon quadratic MPC, short prediction window and high state cost

4.2.1 Short Horizon

In a first simulation, a short sampling period of 1 s and a prediction horizon of ten samples are used, granting only 10 s of prediction, which is a much shorter time frame than that of the orbital motion, given that the orbital period is approximately 96 min. Furthermore, a high state cost is used, which results in the whole manoeuvre being performed in a straight line in approximately 60 s, as shown in Fig. 4.3. The generated manoeuvre requires a ΔV of 6.77 m/s and a thrust that would greatly exceed the capabilities of the small spacecraft. In comparison, the ideal V-bar transfer manoeuvre with two horizontal impulses requires a total ΔV of 3.45 mm/s, as determined with Eq. (3.52), which is almost 2000 times less spent fuel. On the other hand, the ideal manoeuvre requires one orbital period to complete, but because the spacecraft fuel is extremely limited, and in order to take advantage of the relative dynamics, this is the standard time frame for rendezvous operations.

To decrease the value of the manoeuvre ΔV, the state cost is reduced, which makes the controller slower and decreases the fuel spent. However, as observed in Fig. 4.4, the system is now more affected by the natural drift between the satellites and is not able to converge on the reference. This happens because the controller only predicts 10 s ahead, and so is not able to predict much of the natural drift, and the cost function limits the authority to react to it.

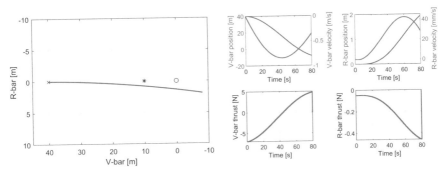

Fig. 4.4 V-bar transfer manoeuvre with receding-horizon quadratic MPC, short prediction window and lower state cost

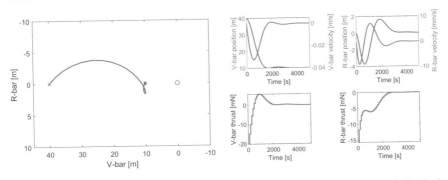

Fig. 4.5 V-bar transfer manoeuvre with receding-horizon quadratic MPC and long prediction horizon

4.2.2 Long Horizon

To counteract the previous drawbacks, the prediction window is increased to the order of magnitude of the orbital period. However, since increasing the prediction horizon significantly worsens the computation time, the sampling period is instead increased to 100 s, granting the controller a prediction of 1000 s. The input cost is also increased, to compensate for the fact that thrust intervals are now longer and so have a bigger effect on the system. As observed in Fig. 4.5, the chaser actuation now has less amplitude, but because its prediction is improved it can use the natural relative motion to better converge on the reference. This feature is also evidenced by the fact that now there is extensive R-bar actuation, despite the fact that the reference is only offset in V-bar from the initial position. The manoeuvre ΔV is now 111 mm/s, which is a significant improvement, but still far from the ideal figure.

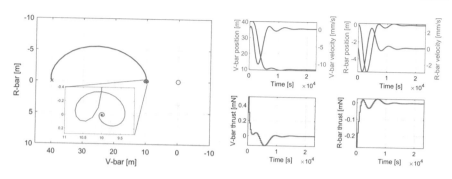

Fig. 4.6 V-bar transfer manoeuvre with receding-horizon quadratic MPC, long prediction horizon and no Lagrangian state cost

4.2.3 Zero State Lagrangian

To improve the performance, the state Lagrangian cost can be disregarded ($Q =$ 0), which allows the controller to better plan ahead, since now it is not penalized for not being on the reference state while being halfway through the manoeuvre. Furthermore, in an attempt to reproduce the ideal V-bar transfer manoeuvre, the sampling time is chosen such that the prediction window is exactly one orbital period. The prediction horizon N is also increased in order to avoid the sampling time becoming too great.

As shown in Fig. 4.6, the trajectory resembles that of the ideal manoeuvre, although the spacecraft cannot reach the reference in one orbital period, and in the final approach it circles the reference, getting closer and closer while never reaching it. This is due to the receding-horizon strategy; since the prediction horizon slides forward every sample, the state that is being tracked is always one orbital period away, and so the controller never makes the final effort to reach the reference. Therefore, a different strategy is required.

4.3 Fixed-Horizon Model Predictive Control

An alternative to the receding-horizon strategy is to decrement the prediction horizon for every sample, such that its edge is always at the same time instant, a procedure known as Fixed-Horizon MPC (FH-MPC) [3]. This, together with the use of a terminal state cost and no Lagrangian state cost, allows for the manoeuvre to be completed in a specified amount of time, which in turn allows for the controller to generate the ideal optimal manoeuvres. The parameters for the following experiments are presented in Table 4.2. Since the prediction horizon is decremented, the computational complexity decreases every sample, and therefore the worst case is now shown instead of the average.

Table 4.2 Controller parameters and results for V-bar transfer manoeuvre simulations with FH-MPC

Figures	T_s (s)	N	R	Q	Q_f	u_{max}	ΔV (mm/s)	t_{max}
4.7	290	20	I	0	$100I$	–	7.61	121 µs
4.8	290	21	–	–	–	–	3.45	7.97 ms
4.9	58.0	100	–	–	–	1 mN	4.68	9.40 ms

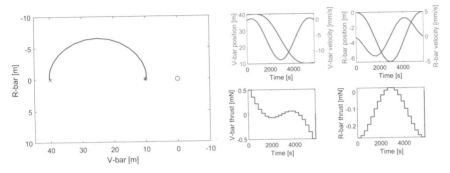

Fig. 4.7 V-bar transfer manoeuvre with FH-MPC and quadratic input cost

With the same conditions as in Fig. 4.6 and the FH strategy yields the result in Fig. 4.7. The trajectory closely resembles that of the ideal manoeuvre and the system reaches the reference in exactly one orbital period, although the ΔV applied is still over two times greater. This is due to the quadratic cost used for the input variable, meaning that the cost function is not directly proportional to the manoeuvre ΔV, since this parameter is linearly proportional to the absolute value of the input force. This also results in a continuous control action instead of sparse like the ideal manoeuvre, with just two thruster burns.

4.3.1 Fuel-Optimal LP Formulation

To obtain a fuel-optimal trajectory, the ℓ_1-norm must be used for the input cost instead of the quadratic function, while the terminal state cost remains quadratic. This resembles the LASSO cost function discussed in Sect. 2.2.4, with $R = 0$ and $R_\lambda = I$. However, the inclusion of the ℓ_1-norm term significantly increases the complexity of the optimization problem, since it is no longer a QP.

One possible way to simplify the optimization problem is to modify the prediction model such that the input forces are split into its positive and negative components

$$F = F^+ - F^-,\qquad(4.2)$$

by extending the B matrix, and thus extending the input vector to

$$u = \left[F_x^+, F_x^-, F_y^+, F_y^-, F_z^+, F_z^- \right]^\top. \tag{4.3}$$

This increases the number of optimization variables, which is disadvantageous, but now each input variable can only take positive values, which makes its absolute value equal to itself. Thus, the ℓ_1-norm can be discarded and the formulation becomes

$$\min_{\substack{\bar{u}_0,\dots,\bar{u}_{N-1} \\ \bar{x}_0,\dots,\bar{x}_N}} \quad (\bar{x}_N - x_{ref})^\top Q_f (\bar{x}_N - x_{ref}) + \sum_{i=0}^{N-1} \Delta t_i \mathbf{1}^\top \bar{u}_i, \tag{4.4a}$$

$$\text{s.t.} \quad \bar{x}_0 = x_t, \tag{4.4b}$$

$$\bar{x}_{k+1} = A_k^{k+1} \bar{x}_k + B_k^{k+1} \bar{u}_k, \quad k = 0, \dots, N, \tag{4.4c}$$

$$\bar{u}_k \geq 0, \quad k = 0, \dots, N-1, \tag{4.4d}$$

which is again a QP, where $\mathbf{1}$ is a column vector of 1's. Note that the input variables are weighed by the time interval between samples Δt, which is crucial in obtaining a fuel-optimal formulation in case the dynamics are not sampled with constant time intervals, since the ΔV is proportional to the duration of the thrusters are fired. Furthermore, constraint (4.4d) ensures that this formulation is equivalent to that with the ℓ_1-norm since it constrains each of the input force components to be equal to or greater than zero.

The formulation can be further simplified by using a terminal state constraint, instead of a terminal state cost, and thus the problem becomes

$$\min_{\substack{\bar{u}_0,\dots,\bar{u}_{N-1} \\ \bar{x}_0,\dots,\bar{x}_N}} \quad \sum_{i=0}^{N-1} \Delta t_i \mathbf{1}^\top \bar{u}_i, \tag{4.5a}$$

$$\text{s.t.} \quad \bar{x}_0 = x_t, \tag{4.5b}$$

$$\bar{x}_{k+1} = A_k^{k+1} \bar{x}_k + B_k^{k+1} \bar{u}_k, \quad k = 0, \dots, N, \tag{4.5c}$$

$$\bar{u}_k \geq 0, \quad k = 0, \dots, N-1, \tag{4.5d}$$

$$\bar{x}_N = x_{ref}, \tag{4.5e}$$

which is now a Linear Program (LP) and can be solved very efficiently. Another advantage of this formulation is that, once the manoeuvre duration is defined, there are no controller parameters that need to be tuned and, in the absence of disturbances, the controller will always reach the reference state without static error. Thus, since the cost function only contains one term that is linearly proportional to the ΔV and since the optimization problem is convex, this formulation is guaranteed to always generate the fuel-optimal trajectory in the specified transfer duration. The disadvantage of using a hard terminal constraint is that the optimization problem

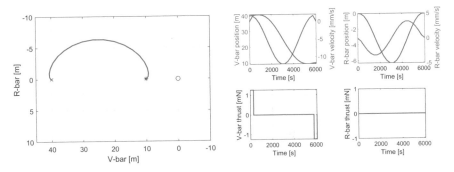

Fig. 4.8 V-bar transfer manoeuvre with fuel-optimal linear FH-MPC

may become infeasible, given the constraints and the length of prediction horizon, especially since in the FH-MPC strategy the prediction horizon is decremented until it is only one sample. However, in the absence of perturbations and disturbances, the problem will never become infeasible if previous iterations are feasible, and so this issue will only be addressed in Sect. 4.6.

Figure 4.8 shows the result of the fuel-optimal linear FH-MPC formulation applied to the one-orbit V-bar transfer. It can be observed that the manoeuvre is performed solely with two thruster actions in the horizontal direction, much like the ideal impulsive manoeuvre in Fig. 3.11. Furthermore, the total ΔV applied is approximately the same as in the ideal manoeuvre, validating the fact that this formulation is fuel-optimal. Note that, to obtain this exact ΔV value, the manoeuvre duration has to be one sample more than the orbital period, to account for the time it takes to perform the last braking input action.

4.3.2 Control Saturation

Limiting the maximum thrust that can be applied by each of the spacecraft thrusters, constraint (4.5d) becomes

$$u_{max} \geq \bar{u}_k \geq 0, \qquad (4.6)$$

where $u_{max} \in \mathbb{R}^6$ is the maximum thrust for each of the input components. Notice that with this constraint the maximum thrust is independent for each direction. If the spacecraft does not have omnidirectional thrust, however, it is more appropriate to constrain the magnitude of the total thrust vector, although this cannot be performed with linear constraints. In [33], this more realistic constraint is utilized, formulating the optimization problem as a second-order cone program. However, if the simpler constraint is used, as it will be in this work in order to maintain the optimization problem as an LP, u_{max} must be lesser than the physical maximum thrust possible, which is a suboptimal approach since the full capacity of the thrusters is not utilized [4]. In

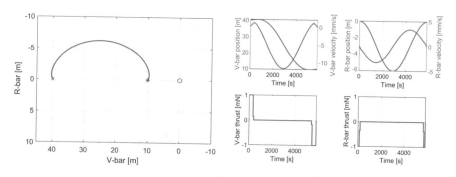

Fig. 4.9 V-bar transfer manoeuvre with fuel-optimal linear FH-MPC, control limits and increased prediction horizon

this work, we assume the chaser spacecraft propulsion system has omnidirectional thrust, as is often the case, and that its attitude relative to LVLH remains constant, and thus constraint (4.6) realistically models the spacecraft thruster limitations.

In Fig. 4.9, the prediction horizon is increased and control limits of 1 mN for each component are added, and thus it can be observed that the initial and final control actions now happen over more than one sample. Also, the ΔV has slightly increased from the previous experiment, which is due to the fact that the thrusters are now limited. Finally, from Table 4.2, it can be seen that, despite having increased the prediction horizon by a factor of five and adding more constraints, the computation time only increased by approximately 18%, which is due to the fact that the optimization problem is an LP and can be solved very efficiently. Hence, we showed that this formulation can generate fuel-optimal trajectories, similar to the ones already used to plan rendezvous missions and that it can generate them very efficiently and feasibly in real-time.

4.4 Variable-Horizon Model Predictive Control

The previous formulation optimizes the fuel required for the manoeuvre given a pre-specified transfer duration. However, it is also desirable to optimize the manoeuvre duration as well as fuel expenditure. This requires adding the prediction horizon N as an integer optimization variable, and so the formulation known as Variable-Horizon MPC (VH-MPC) becomes

$$\min_{\substack{\bar{u}_0,\ldots,\bar{u}_{N_{max}-1} \\ \bar{x}_0,\ldots,\bar{x}_{N_{max}} \\ N \in \mathbb{N}}} \gamma N + \sum_{i=0}^{N_{max}-1} \Delta t_i \mathbf{1}^\top \bar{u}_i, \tag{4.7a}$$

$$\text{s.t.} \quad \bar{x}_0 = x_t, \tag{4.7b}$$

$$\bar{x}_{k+1} = A_k^{k+1} \bar{x}_k + B_k^{k+1} \bar{u}_k, \tag{4.7c}$$

$$0 \le \bar{u}_k \le u_{max}, \ k = 0, \ldots, N_{max} - 1, \tag{4.7d}$$

$$\bar{x}_N = x_{ref}, \tag{4.7e}$$

$$1 \le N \le N_{max}, \tag{4.7f}$$

where N_{max} is the bound for the prediction horizon. The prediction horizon is also added to the cost function, and thus the parameter γ is used to tune the trade-off between transfer time and fuel consumption. If $\gamma = 0$, the solution is the manoeuvre duration that minimizes the fuel within the bounds of the prediction horizon; for simple manoeuvres in a circular target orbit, the solution will typically be at N_{max}, while for more complex manoeuvres and in elliptic target orbits that may not be the case. Constraint (4.7e) now becomes nonlinear, since it indexes an optimization variable with another, and thus the problem is a mixed-integer nonlinear program, which is computationally expensive to solve.

A method of transforming this problem into a Mixed-Integer Linear Program (MILP) was first presented by Richards and How [34] and applied to rendezvous in [3], requiring the substitution of the prediction horizon by two vectors of binary optimization variables. Variable v_k is 1 if the manoeuvre is completed exactly at instant k, and 0 otherwise, and thus

$$\sum_{k=1}^{N_{max}} v_k = 1. \tag{4.8}$$

The variable p_k is 1 while the manoeuvre is not completed, and 0 afterwards, and so conceptually, we have

$$\sum_{k=1}^{N_{max}} p_k = N. \tag{4.9}$$

The two binary variables are related by the dynamic equation

$$p_{k+1} = p_k - v_{k+1}, \tag{4.10}$$

that maintains their integrity: if the manoeuvre is completed at $k + 1$, then $v_{k+1} = 1$ which forces p_{k+1} to flip values, otherwise, it is maintained. The VH-MPC MILP formulation then becomes

$$\min_{\substack{\bar{u}_0,\ldots,\bar{u}_{N_{max}-1} \\ \bar{x}_0,\ldots,\bar{x}_{N_{max}} \\ p_0,\ldots,p_{N_{max}}\in\{0,1\} \\ v_1,\ldots,v_{N_{max}}\in\{0,1\}}} \gamma \sum_{i=0}^{N_{max}} p_i + \sum_{i=0}^{N_{max}-1} \Delta t_i \mathbf{1}^\top \bar{u}_i, \tag{4.11a}$$

$$\text{s.t.} \quad \bar{x}_0 = x_t, \tag{4.11b}$$

$$\bar{x}_{k+1} = A_k^{k+1}\bar{x}_k + B_k^{k+1}\bar{u}_k, \tag{4.11c}$$

$$0 \le \bar{u}_k \le u_{max}, \quad k = 0, \ldots, N_{max} - 1, \tag{4.11d}$$

$$-(1 - v_k)h \le x_k - x_{ref} \le (1 - v_k)h, \tag{4.11e}$$

$$p_{k+1} = p_k - v_{k+1}, \quad k = 0, \ldots, N_{max} - 1, \tag{4.11f}$$

$$p_{N_{max}} = 0, \tag{4.11g}$$

$$\sum_{k=1}^{N_{max}} v_k = 1. \tag{4.11h}$$

Notice that the optimal prediction horizon is no longer implicitly included in the cost function, but rather the sum of the p variables, since we have the relation in (4.9). Also, the properties in Eqs. (4.8) and (4.10) are included as optimization constraints in (4.11h) and (4.11f), respectively, while the constraint (4.11g) forces the manoeuvre to be completed at least by the end of the maximum prediction horizon. Lastly, the terminal state constraint in (4.11e) is now an inequality constraint, where parameter h is a sufficiently large number. Thus, the term $(1 - v_k)$ allows to trigger the terminal constraint: at the moment the manoeuvre is completed, v_k is 1 and this term is 0, such that the bounds of the linear inequality are tight and the terminal state constraint becomes active; otherwise, it is 0, and so the bounds are very wide and thus the constraint becomes inactive.

The computational load for this formulation is bigger than for the FH-MPC since MILP problems are harder to solve. Thus, in a real-time scenario, it may be preferable to predetermine offline the manoeuvre duration and use the FH-MPC formulation instead. However, as proposed in [10], determining the manoeuvre duration offline can be performed in an optimal way by using the VH-MPC formulation. The optimal transfer time may change slightly along the way due to disturbances, but the time determined offline can still be expected to remain approximately optimal. As will be discussed in Sect. 4.6, however, using the VH-MPC formulation online is advantageous for its feasible robustness property. When used online, if the maximum manoeuvre duration counting from the initial instant is to be maintained for subsequent iterations then N_{max} should be decremented every instant; otherwise, the controller might deviate from its initial trajectory and extend the manoeuvre beyond the initial maximum final instant in order to spend less fuel.

In [10], the authors present an extension to the VH-MPC framework that allows for multi-step manoeuvres to be considered, where the durations of the sub-manoeuvres are optimized simultaneously and offline, also via integer linear programming. The resulting multi-step manoeuvre is then performed online as a sequence of several FH-MPC manoeuvres.

4.5 Passive Safety

It is a requirement in rendezvous mission design that, besides ensuring that the nominal trajectory does not cause a collision between the spacecraft, the free-drift motions from any point in the trajectory also remain collision-free, within a specified time horizon. Designing trajectories in this way ensures that in the case that the thruster fails to fire, the two spacecraft will not collide due to the natural drift, which is thus is known as *passive safety*. Furthermore, in the event of any other fault that warrants an abort in the approach, it becomes a safe strategy to simply shut down the spacecraft thrusters. Figure 4.10 illustrates the need for designing passively safe rendezvous trajectories, where a V-bar transfer manoeuvre is shown such that a collision occurs after one orbit in case the final thrust fails.

Typically, passively safe trajectory design is performed by choosing specific classes of manoeuvres with good passive safety properties. For example, in the example in Fig. 4.10, if radial pulses are applied instead of horizontal, the nominal trajectory will be similar to the one in Fig. 3.10, and the free-drift failure trajectory would be similar to Fig. 3.6, which returns to the initial position after half an orbit, thus avoiding collision in an infinite horizon. Ensuring passively safe trajectories usually comes at the cost of increased fuel expenditure; in the previous example, the radial-pulse manoeuvre requires a ΔV over four times greater.

For rendezvous with MPC, passive safety design can be included in the optimization problem as a constraint. Typically, this is performed by constraining the discrete states in the nominal and failure trajectories to be outside of the target spacecraft or

Fig. 4.10 Illustration of the passive safety problem in a V-bar transfer manoeuvre

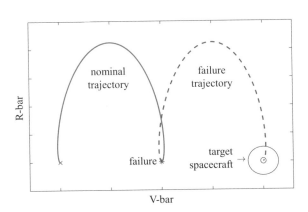

its safety region. Thus, each different failure trajectory will have a separate obstacle constraint. To simplify, we will first detail how to formulate a general obstacle avoidance constraint for the nominal trajectory, and then extend it to the failure trajectories in order to formulate the passive safety constraint.

4.5.1 Obstacle Avoidance with Nonlinear Optimization

The most straightforward approach to formulating an obstacle avoidance constraint is to simply constrain the states to be outside the region defined by the obstacle. Since this constrains a connected region of the state space, the feasible set becomes nonconvex. Thus, these obstacle avoidance constraints result in an optimization problem with nonlinear constraints, which makes it difficult to solve and introduces different local minima.

For example, for the circular obstacle illustrated in Fig. 4.11, the obstacle avoidance constraint becomes

$$\|x_k - c\|^2 \geq r^2, \ k = 1, \ldots, N, \tag{4.12}$$

requiring N nonlinear optimization constraints. This problem has been previously solved in the literature with Sequential Quadratic Programming (SQP) algorithms [35, 36]. However, with the passive safety constraints as well, the number of nonlinear optimization constraints increases dramatically, greatly affecting computational performance and introducing significant non-convexities, thus likely making it infeasible to use this approach in real-time.

An alternative method of formulating an obstacle avoidance constraint is with the use of linear obstacle constraints and auxiliary binary optimization variables such that only one of the constraints is active at each time, thus turning the problem into a MILP [37]. Although a MILP can have better properties, it cannot be solved in

Fig. 4.11 Illustration of obstacle avoidance of circular object with nonlinear constraints

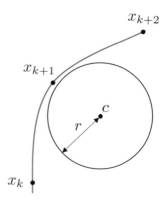

polynomial time, and the complexity drastically increases with the inclusion of the failure trajectories, and therefore it may also not be feasible to implement in real-time.

4.5.2 Obstacle Avoidance with Linear Optimization

Another method to perform obstacle avoidance is with pure linear optimization. In [38], the obstacle constraint is replaced with a convex set that excludes the obstacle but is static along with the prediction. This approach does not allow for trajectories that bend around the obstacle and otherwise results in a significant ΔV increase, since the trajectory must lie within a much more conservative region, often even leading to infeasibility.

Another approach is to subject each state in the trajectory to a different linear inequality constraint

$$D_k x_k \leq b_k, \; k = 1, \ldots, N, \tag{4.13}$$

which is tangent to the original obstacle, as illustrated in Fig. 4.12. The linear constraints have to be determined prior to the optimization. Furthermore, the state at each time is now subject to a more conservative constraint, which can affect the optimality of the trajectory. On the other hand, the optimization problem becomes an LP again, allowing it to be solved much more efficiently.

The method used to determine the linear constraints $D_k x_k \leq b_k$ currently found in the literature is to rotate the constraint around the obstacle with time [4, 5, 8, 39], where the rate at which it rotates is a controller parameter that must be optimized offline. However, this method is not appropriate to use in the passive safety problem, as each failure trajectory would require different rates of rotation that must be optimized simultaneously.

On the other hand, the linear constraints completely cover the obstacle, meaning that there is no possibility of the original constraint being violated in the optimization. It can, however, be violated after a disturbance acts on the system, an issue that will be addressed in Sect. 4.6.

Fig. 4.12 Illustration of obstacle avoidance with linear constraints

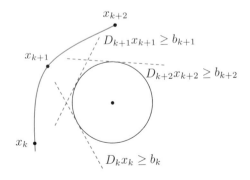

4.5.3   Obstacle Avoidance with Sequential Linear Programming

In this book, we propose a different method for satisfying obstacle avoidance constraints, based on the previous, but which is easier to apply since it does not require parameter tuning. It does, however, rely on a sequence of LPs solved online, as opposed to just one.

First, the problem without any obstacle avoidance constraints is solved, where, naturally, violations of the obstacle avoidance constraint can occur. Second, the planes tangent to the obstacle and facing each of the discrete states in the trajectory are determined, as shown in Fig. 4.13a. Then, the tangent planes are used as linear constraints for a second LP optimization, which yields the result in Fig. 4.13b. The trajectory now avoids the collision but is more conservative than necessary since it flies by the obstacle at a distance. The process can then be repeated, where this trajectory is used to again determine the linear constraints for another LP optimization. This procedure yields the trajectory in Fig. 4.13c, which is very similar to that which is be obtained with nonlinear optimization.

It is thus possible to achieve obstacle avoidance with a sequence of purely linear optimizations. However, it is remarked that, in some situations, the linear constraints determined from the first unconstrained optimization can render the feasible region empty, as will be shown in Sect. 4.7.3, due to the presence of a terminal hard constraint. This drawback, however, can be overcome with, for example, virtual control techniques [40], although this is not addressed in this work. Furthermore, there is currently no guarantee that the trajectory converges to a local minimum of the problem with the nonlinear obstacle constraints, which will be the subject of future work. The possibility for the LP to be unbounded is also of concern, although intuitively this does not happen since the underlying problem without the linearised obstacle constraints is bounded; formal proof of this will also be the subject of future work.

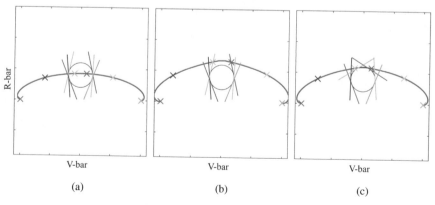

Fig. 4.13 Illustration of the method for determining the linear obstacle avoidance constraints with iterative optimization with linear constraints

This technique will hereby be referred to as *Obstacle Avoidance with Sequential Linear Programming* (OASLP) and Algorithm 4.1 summarizes the strategy, where the convergence criteria for stopping may be the trajectory change between iterations being within a specified tolerance. It can be feasible to use the OASLP technique online since it relies purely on linear optimization.

Algorithm 4.1 Obstacle Avoidance with Sequential Linear Programming

---

1 Solve optimization problem without obstacle constraints
2 repeat
3 | Determine planes tangent to obstacle facing each point in the trajectory
4 | Solve optimization problem with tangent planes as linear constraints
5 until *trajectory convergence*

The concept behind this algorithm is similar to that of a typical Sequential Linear Programming (SLP) and SQP algorithms [41] for nonlinear programming, in that each sub-problem is a convex approximation of the original one. The main difference is that for the present method each linearized sub-problem optimizes the original problem variables, as opposed to SLP and SQP algorithms optimizing at each iterate the search direction within a trust region. Thus, the OASLP method is more comparable to Successive Convexification [40] or Sequential Convex Programming [42] algorithms for optimal control, with a first-order constraint approximation, specific for obstacle avoidance constraints. Given that the original LP without the obstacle constraints is guaranteed to be bounded, there is no need for a trust region, which the mentioned methods utilize in order to prevent unboundedness.

4.5.4 *Passive Safety Constraint*

To formulate the rendezvous passive safety constraint, the failure trajectories must be propagated with the prediction model, as first proposed by Breger and How [38], and then constrained with any of the methods presented in the previous sections. If a total thruster failure occurs at time k, the resulting free-drift failure trajectory x_{F_k} is described by

$$x_{F_{k,t}} = A_k^t x_k, \quad t > k, \tag{4.14}$$

where A_k^t is the dynamic matrix that transitions the state from instant k to t, as illustrated in Fig. 4.14.

The passive safety constraint then becomes

$$x_{F_{k,t}} \notin Obstacle, \quad k \in \{1, \ldots, N\},$$
$$t \in \{k+1, \ldots, k+S\}, \tag{4.15}$$

Fig. 4.14 Propagation of the
free-drift failure trajectories

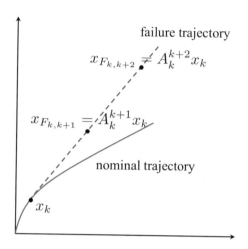

where failures at all discrete instants in the trajectory are considered and tracked for S samples, where S is the safety horizon. This can be seen as a type of move blocking strategy, as described in Sect. 2.4, where the control actions past the control horizon are zero.

Notice from (4.15) that, while optimizing at instant $k = 1$, the failure trajectory for instant $k = N$ is considered and constrained, which might seem conservative; this approach, however, allows the controller to generate a more accurate trajectory right away and avoid corrections later. However, this method requires $N \times S$ optimization constraints, which have a great computational burden and thus require an efficient implementation, such as the OASLP method presented in Sect. 4.5.3. If this method is utilized, each failure trajectory will have its own set of linear constraints. Furthermore, an additional N constraints are required for the nominal trajectory collision avoidance. One possible way to reduce the online complexity is to check at each optimization if any failure trajectory is far from violating the constraint, or superimposed with other trajectories due to inaction at some points, and remove the constraints associated to those for the next optimization. This strategy is not employed here, but it will be explored in future work.

Note that this method does not consider thruster failures during a ΔV, assuming only that the thrusters fail to fire in the first place. Often those types of failures will also be covered as a consequence, although this cannot be guaranteed. Because MPC operates in discrete time, it is not possible to consider mid-thrust failures at every continuous-time instant, although some additional discrete intermediate points can be considered, at the cost of greater computational complexity.

Another disadvantage of working in discrete time is that only the discrete states are constrained to be outside the obstacle, and not the whole continuous trajectory, which can lead to a collision if the time between samples is too great. This can be minimized by decreasing the sampling time or by including extra intermediate samples, both at the cost of a greater computation time. An alternative approach presented in [43] allows to constrain trajectories in continuous time, by transforming the optimization problem into a semidefinite program. This approach also comes at the cost of increased computation time, but completely eliminates the issue.

Finally, another limitation of this method is that it only guarantees passive safety within a finite horizon S. For most operations, this is sufficient since it gives ground operators enough time to react accordingly in face of a fault. Sometimes, however, it might be desirable to achieve passive safety with an infinite horizon. In [38] this is achieved by forcing all failure orbits to be invariant with respect to the target, via the constraint

$$x_{F_{k,k}} = A_k^{k+N_o} x_{F_{k,k}}, \quad k \in \{1, \ldots, N\}, \tag{4.16}$$

where N_o is the number of samples in an orbit. This, however, can constrain the problem too much and easily make it infeasible, and in [38] it is only tested for very close-range operations and in a circular target orbit. Another interesting problem regarding this topic is preventing formation evaporation, meaning that the failure trajectories are optimized such as to prevent the spacecraft from drifting too far apart.

4.6 Robust Rendezvous

There are many sources of disturbances in a real rendezvous mission scenario, with respect to which the guidance and control (G&C) system must be robust. Firstly, there are modelling errors, since the prediction model used is a linearization of the real dynamics. Furthermore, this approximation emerged from a nonlinear model that assumed the gravitational field of the central body is uniform, which is never the case and is another significant perturbation on the model, of which the most significant for Earth satellites is the J_2 effect, due to the planet's oblateness.

The relative navigation sensors and algorithms introduce errors on the position of the chaser relative to the target, and navigation uncertainty in the absolute position of the spacecraft is also present, which generates errors in the orbital parameters used for the prediction model. Another considerable perturbation is actuator errors, which are in magnitude, due to imprecise thruster action, in direction, due to error in the spacecraft orientation and thrusters mounting misalignment, and in burn duration, due to imprecise timing. Often, spacecraft thrusters can only be turned on or off, with no intermediate thrust possible. Thus, intermediate thrust commands are performed via *pulse width modulation* (PWM), spacecraft, which is another source of mismatch between prediction and reality unless modelled. Furthermore, there is also a minimum thrust value that can be generated via PWM, due to the delay in opening the thruster valves, which may be another source of modelling errors.

Finally, there are undesirable external forces acting on the spacecraft which are another source of disturbance, such as atmospheric drag, solar radiation pressure, or the gravitational effect of other massive bodies, such as the Moon. These, however, affect both spacecraft and thus only the difference in forces generate a disturbance in the relative position, although absolute forces will change the orbit over time.

Robustness to these disturbances means not only that the system remains stable but also that it can still accurately converge to the reference, in approximately the specified manoeuvre duration and without a very significant increase in ΔV, in what we address here as *robust performance*. Furthermore, it is necessary that state constraints, such as passive safety, are not violated due to disturbances, which is known as *robust constraint satisfaction*. Finally, the disturbances can often push the system to a state that renders the optimal control problem infeasible, and thus the controller must have *robust feasibility*.

There is some inherent stability robustness for MPC, resulting from the fact that it is a closed-loop control strategy [44]. Sometimes this is sufficient, although often robust strategies must be employed to increase performance and to ensure state constraints are not violated due to a disturbance. Several robust strategies exist for general Robust MPC, such as Min–Max Feedback MPC [45] and Tube MPC [46], although it is often difficult to feasibly implement these in real time. More on Robust MPC can be found in Rawlings et al. [44]. We will first review some of the robustness strategies presented in the literature specifically for MPC for rendezvous, before presenting our own contributions.

4.6.1 Robustness Techniques Review

In an early robustness technique for rendezvous, [47] extends the LP formulation and considers uncertainty in the initial condition, optimizing the trajectory simultaneously for multiple initial states. The technique is tested with nonlinear simulations that include differential drag, and although it is an improvement upon the non-robust controller, it is overly conservative and requires a ΔV much greater than in the unperturbed case. Also, no constraints at all are considered.

A different approach is taken in [3], extending the VH-MPC formulation to yield robustness in face of an arbitrary unknown but bounded disturbance. Using the VH strategy solves the terminal constraint infeasibility that can easily occur for FH-MPC at the end of the manoeuvre when the prediction horizon is short. Furthermore, the VH formulation is extended with correction control decisions in the next two steps which are limited given the disturbance bounds. This two-step correction of the disturbance guarantees that if the problem is feasible at the current step, it will also be at the next, thus achieving robust feasibility. Furthermore, the correction steps ensure that, given any disturbance, the state can converge to the nominal trajectory in two steps, hence granting robust stability. The downsides for this approach are that only bounded additive disturbances are considered, and a MILP formulation is required which can be difficult to implement in real time. Also, the robust strategy required almost twice the nominal fuel for one simulation, and it does not address robust constraint satisfaction.

In [48], a bounded but unknown navigation error is considered, and a technique based on the feedback-MPC strategy from the Tube-MPC framework is presented, meaning that the decision variables are feedback policies, parametrized as affine control laws, instead of control actions. The state uncertainty at each step is propagated as elliptical sets, and the optimal control problem becomes the minimization of the terminal uncertainty set, turning it into a convex conic optimization problem. The advantage in the use of feedback-MPC is that the full plan can be determined offline, while the online work becomes the computation of small disturbance correction terms, which is performed with simple algebraic computations. However, the fuel optimality of the resulting controller is not evaluated, and more complex disturbances for which the computation of bounds is not possible are not considered. Furthermore, only control saturation constraints are included, and thus robust state constraint satisfaction is not addressed. Other robustness approaches also based on Tube-MPC can also be found in [49, 50].

In [51], the same authors consider bounded execution errors in firing time and orientation. The terminal state constraint is substituted by a convex polytopic set, where its dimensions are minimized as decision variables such that it is guaranteed to contain the terminal state, taking into account the propagation of the uncertainty due to the bounded errors. By assuming the terminal polytopes are parallelotopes, linearizing the effect of orientation errors, and introducing several auxiliary optimization variables, the problem can be formulated as an LP. A fuel budget constraint is included to limit propellant consumption, but because the cost function contains only the dimensions of the terminal set, the formulation is not necessarily fuel-optimal. Furthermore, magnitude thruster errors are not considered, although simulations are performed with the nonlinear dynamics including the J_2 effect and atmospheric drag. Navigation errors are also not considered, and this method is incompatible with the one in [48] due to necessarily very different formulations.

An approach that simply relies on the intrinsic robustness of MPC is presented in [4] for in-plane proximity manoeuvring, using a quadratic cost and the Receding-Horizon strategy. The controller is shown to be robust for large actuation disturbances of up to $\pm 25\%$ magnitude and $\pm 45°$ in orientation and to solar pressure and atmospheric drag. However, the authors do not analyse fuel consumption, although it can be seen that it is significantly greater than the ideal sparse manoeuvres since the resulting trajectories are approximately straight-line approaches with significant non-sparse actuation. Furthermore, navigation errors are not considered and robust constraint satisfaction is not addressed. This work is extended by [8], by including the out-of-plane motion, obstacle constraints, and formulating two different controllers for the rendezvous and docking phases of the mission, for which the first is complemented with a reference governor. The manoeuvres generated appear to require a ΔV orders of magnitude higher than is typically expected for these medium distance manoeuvres in a circular orbit. We strongly believe that a robust MPC controller for rendezvous should be based on the Finite-Horizon strategy with a cost function proportional to the ΔV, as was presented in Sect. 4.3.

An effective method for robust constraint satisfaction is *constraint tightening* [52], in which the bounds of state constraints are constricted along the horizon, in order to account for disturbances that could push the system to the infeasible region. The greatest advantage of this approach is that the complexity of the original problem is retained. However, the tightened constraints can become too conservative, affecting performance and causing infeasibility. This technique is used in a rendezvous application by Breger et al. [6], where the tightened constraints are determined by propagating uncertainty sets with a preselected feedback nilpotent control law and given a bounded navigation disturbance and computing the Pontryagin difference between the original constraint region and the uncertainty region. However, this method is not applicable when dynamics are time-varying, and is applicable with unbounded disturbances.

References [7, 32] present a probabilistic constraint tightening approach, labelled chance-constrained MPC, where all disturbances are assumed to be Gaussian and are estimated in real time. The constraint bounds are then adjusted taking into account the uncertainty due to disturbances, guaranteeing that the original constraints are satisfied with a specified probability. Although many disturbances are not additive in nature, this approach shows good performance when simulated with the nonlinear dynamics, thrust magnitude and orientation errors, and unmodeled eccentricity, where the state constraint considered is a line-of-sight cone. However, only circular orbit dynamics are considered, and the issue of infeasibility when constraints become too conservative is not addressed. This technique is extended for a general elliptic orbit in [10] and also improved by exploiting the fact that navigation error Gaussian parameters are often known and don't need to be estimated. Furthermore, infeasibility in the terminal state constraint is dealt with a technique first presented in [53], where the terminal constraint is allowed to be relaxed into a terminal box in face of infeasibility. The technique presents a good performance in the presence of nonlinear dynamics, J_2 effect, atmospheric drag, navigation errors and thruster errors, although the parameters chosen for these last two are quite small and not representative of real conditions, such as those in the PROBA-3 mission.

As can be seen, no single robustness strategy has yet emerged as the definite standard approach. These often neglect one or more of the core requirements for a robust MPC controller for rendezvous, such as fuel optimality, being computationally feasible for real-time operation, manoeuvre accuracy, guaranteeing robust constraint satisfaction and robust feasibility, or being robust to all possible types of disturbances and uncertainty. The chance-constrained approach in [7, 10] is a good candidate, since it takes into account all disturbances via online estimation in order to ensure constraint satisfaction without increasing the complexity of the optimization problem, uses the fuel-optimal FH-MPC formulation, and deals with infeasibility in the terminal state constraint. However, as will be shown later on, in the presence of navigation and thruster errors in the magnitudes expected for the PROBA-3 mission, the FH-MPC strategy faces some difficulty related to fuel expenditure and reference convergence. Furthermore, the constriant-tightening approach has not been tested with passive safety constraints, although with the obstacle avoidance techniques presented in Sect. 4.5.2 these can be written as linear constraints, to which the chance-constrained technique can easily be applied.

4.6.2 Feasible Terminal Box

In the presence of disturbances, the optimal control problem can often become infeasible, which should be avoided at all costs in a real-time application. The terminal state constraint in the FH-MPC strategy in particular is prone to this issue since its prediction horizon is decremented until only one control move is available. Since the system is not one-step controllable, meaning that it is not possible to transfer between any two arbitrary states in just one control action, a disturbance may push it to a position from which the terminal state cannot be reached in one step, rendering the optimization problem infeasible. This limitation can be further increased by other control and state constraints, and even more so when constraint tightening techniques are used.

One possible solution to the infeasibility problem is to use a quadratic terminal state cost as in (4.4), instead of the terminal constraint. This is undesirable, however, since it turns the optimization problem into a QP, and introduces more parameters for tuning. Another solution that is commonly used is to relax the equality constraint into an inequality and adding margins, thus introducing the concept of *terminal box* [3, 5, 38, 43]. Constraint (4.5e) then becomes

$$- \delta_{box} \leq \bar{x}_N - x_{ref} \leq \delta_{box}, \tag{4.17}$$

where $\delta_{box} \in \mathbb{R}^6$ defines the bounds for the box. The issue with this approach is that to improve the guarantee of feasibility, the size of the terminal box has to be increased, which in turn worsens the accuracy of the system since it will tend to aim for the edges of the terminal box. Furthermore, there is no guarantee that the chosen dimensions for the terminal box will always ensure the existence of a solution for all possible scenarios.

This approach can be improved upon by introducing the dimensions of the terminal box as optimization variables and including them in the cost function for minimization, as proposed in [53]. Introducing the optimization variables $\delta_1, \ldots, \delta_6$, the terminal constraint (4.17) becomes

$$- \begin{bmatrix} \delta_1 \\ \vdots \\ \delta_6 \end{bmatrix} \leq \bar{x}_N - x_{ref} \leq \begin{bmatrix} \delta_1 \\ \vdots \\ \delta_6 \end{bmatrix}, \tag{4.18}$$

which remains a linear constraint with respect to all optimization variables. Furthermore, the cost function (4.5a) now includes the new variables

$$V(\cdot) = \sum_{i=0}^{N-1} \Delta t_i \mathbf{1}^\top \bar{u}_i + \sum_{j=1}^{6} h_j \delta_j, \tag{4.19}$$

where h_j is a large enough number to ensure the controller only relaxes the terminal constraint to ensure feasibility, and not to save fuel. Thus, the terminal box will always have the minimum size that guarantees feasibility, and thus we designate this technique as *feasible terminal box*. Note that, although the box dimensions can be upper-bounded to avoid the terminal box becoming too large, this should be avoided as it no longer guarantees the terminal constraint will always be feasible. It is also not necessary to lower bound the box dimensions with 0 since negative values are already impossible. The obvious drawback for the feasible terminal box approach is that it requires the addition of six new optimization variables, although this is negligible when compared to the usual problem dimensions.

4.6.3 Dynamic Terminal Box

In the presence of stochastic disturbances, such as navigation or actuation errors, the terminal state constraint will cause the controller to perform frequent trajectory corrections, in an attempt to maintain the predicted terminal state exactly on the reference. However, because the controller acts on imperfect information and because its commands are not perfectly executed, this results in an overcorrection that results in the waste of fuel. Such an effect may be minimized by considering a terminal box constraint instead of a terminal equality constraint, thus loosening the terminal constraint and resulting in fewer trajectory corrections. This is similar to the feasible terminal box technique, although the latter is used for preventing infeasibility as opposed to improving fuel consumption, and the two are not mutually exclusive. However, loosening the terminal constraint results in reduced manoeuvre accuracy. Thus, we propose here that the terminal box is reduced as the manoeuvre is performed in what we designate as *dynamic terminal box*, thus achieving reduced fuel expenditure and maintaining accuracy.

One idea for achieving this is to dynamically change the weights h_j of the feasible terminal box technique, increasing them with time such that the box is increasingly tightened. However, because of the sparsity of the linear cost function, changing the box margins cost only results in either the terminal box having the minimum size that allows the problem to be feasible or the terminal box being completely loosened such that no control input is required. An alternative is to take a quadratic cost for the box margins, although this is undesirable since it makes the optimization problem a QP. Alternatively, the box margins may be tuned directly, prior to the optimization.

We propose here a modification on the feasible terminal box constraint to include a time-varying margin ε_t, where t is the time instant at which the optimization problem is being solved. Thus, constraint (4.18) becomes

$$-\begin{bmatrix} \delta_1 \\ \vdots \\ \delta_6 \end{bmatrix} - \varepsilon_t \leq \bar{x}_N - x_{ref} \leq \begin{bmatrix} \delta_1 \\ \vdots \\ \delta_6 \end{bmatrix} + \varepsilon_t. \tag{4.20}$$

Although this approach can achieve increased performance, it requires significant tuning of the margins for each time instant. In Sect. 4.7.4, we experiment with an initial box ε_0 that decreases linearly with time until it is zero at the final iteration. This may be improved, for example by determining the box dimensions as a function of the uncertainty, and thus warrants further research.

4.6.4 Terminal Quadratic Controller

The sparse thrust profile of the FH-MPC formulation is not appropriate for executing accurate manoeuvres in the presence of stochastic disturbances such as navigation and execution errors. This is due to the fact that crucial ΔV's tend to be performed in one sample only while planning under imperfect state information and with imperfect execution of the ΔV. One of these crucial ΔV's is the final braking thrust that is usually performed at the end of the manoeuvre in order to cancel all relative velocity, which under sparse actuation and in the presence of the mentioned disturbances tends to not be very effective.

Thus, we propose here the use of the following terminal linear–quadratic MPC controller to substitute the last sample of the FH-MPC manoeuvre

$$\min_{\substack{\bar{u}_0,\ldots,\bar{u}_{N_T-1} \\ \bar{x}_0,\ldots,\bar{x}_{N_T}}} (\bar{x}_{N_T} - x_{ref})^\top Q_f (\bar{x}_{N_T} - x_{ref}) + \sum_{i=0}^{N_T-1} \bar{u}_i^\top R \bar{u}_i, \tag{4.21a}$$

$$\text{s.t.} \quad \bar{x}_0 = x_t, \tag{4.21b}$$

$$\bar{x}_{k+1} = A_k^{k+1} \bar{x}_k + B_k^{k+1} \bar{u}_k, \tag{4.21c}$$

$$-u_{max} \leq \bar{u}_k \leq u_{max}, \quad k = 0, \ldots, N_T - 1, \tag{4.21d}$$

where x_t is the state measurement/estimate at the penultimate iteration of the manoeuvre, and N_T is the prediction horizon of the terminal controller. The use of the quadratic cost for the input variable results in a less sparse actuation that, combined with the fact that one control decision is substituted with N_T decisions, allows for more precise execution of the final braking ΔV. To maintain the manoeuvre duration specified initially, the prediction horizon N_T covers the same time window as the last sample of the FH-MPC controller. Furthermore, the FH strategy is also utilized, where the prediction horizon is decremented every sample to ensure the manoeuvre is completed in the specified time, although the terminal state constraint is substituted by a terminal quadratic cost. Finally, no intermediate state cost is used, since it decreases the controller performance, as shown in Sect. 4.2.

Since the prediction horizon for the terminal controller is not required to be very long, it may be feasible to implement it with Explicit MPC, granting computational advantages which are desirable given that the braking manoeuvre is critical. Furthermore, because the terminal controller only has to cancel the relative velocity at the end of a manoeuvre that was planned with the FH-MPC formulation with passive safety constraints, the inclusion of these constraints for this controller may be unnecessary, further increasing the feasibility of implementing it with Explicit MPC.

4.7 Tests and Results

This section features several simulations and experiments with the methods presented along with this chapter. The MPC optimization problems are solved in an interpreted programming environment, where a simplex algorithm [41] is used for linear programming on the FH-MPC formulation. For the VH-MPC formulation, a branch-and-bound algorithm is used to solve the resulting MILP, with the previous simplex algorithm solve the linear sub-problems. Finally, a sequential quadratic programming (SQP) algorithm [41] is used for solving the nonlinear program that arises with the passive safety problem, when not solved with the OASLP algorithm. Because these algorithms do not take advantage of the MPC problem structure, the state-substitution technique presented in Sect. 2.2.2 is always utilized, which was empirically shown to result in better performance.

The computation times for solving these optimization problems are always presented, solved with a 4th Generation 2.4 GHz Intel-i7 Processor. Note that the computation times are not truly representative of an on-board environment, not only due to hardware differences but due to the tests running on interpreted code instead of compiled code, which is typical for embedded spacecraft algorithms and performs significantly better.

For experiments in highly elliptical orbits, the conditions of the ESA PROBA-3 rendezvous experiment mission [17] are considered, with an eccentricity of 0.8111 and a perigee height of 600 km.

4.7.1 Fixed Horizon Model Predictive Control

In this section, several rendezvous experiments are performed with the FH-MPC formulation, recreating several of the thrust manoeuvres presented in Chap. 3 and thus proving this formulation is indeed fuel optimal. The V-bar transfer manoeuvre in one orbital period has already been illustrated in Fig. 4.8. The parameters and results for the following experiments are contained in Table 4.3.

In Fig. 4.15, a V-bar transfer manoeuvre with a duration of half an orbital period is presented. The trajectory and thrust profile generated by FH-MPC exactly matches the well-known V-bar transfer with two radial impulses, such as the one presented in Fig. 3.10, and the obtained ΔV is exactly the same as that computed with (3.51).

Figure 4.16 presents another V-bar transfer, but now with a two-orbit manoeuvre duration. It can be seen that the generated trajectory is much like the V-bar transfer manoeuvre with horizontal thrust, such as presented in Figs. 3.11 and 4.8, but with

Table 4.3 Controller parameters and results for V-bar transfer manoeuvre simulations with FH-MPC

Figures	T_s	E_s	e	θ_0	N	ΔV (mm/s)	t_{max} (ms)	t_{avg} (ms)
4.15	59.2 s	–	0	–	50	16.25	8.84	7.37
4.16	58.3 s	–	0	–	200	3.45	12.1	7.79
4.17	58.6 s	–	0	–	100	5.47	11.6	7.68
4.18	59.2 s	–	0	–	50	10.83	8.51	7.38
4.19	–	3.6°	0.4	0°	100	119.0	9.26	7.41
4.20	–	1.8°	0.8111	180°	200	52.1	10.3	8.03
4.21	–	1.70°	0.8111	179°	100	407.4	9.52	7.89

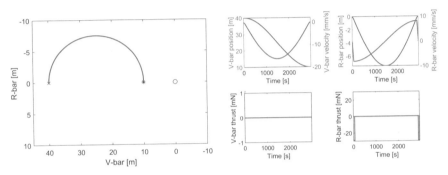

Fig. 4.15 V-bar transfer manoeuvre in half an orbital period with FH-MPC

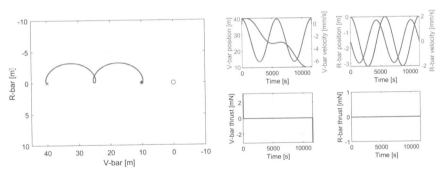

Fig. 4.16 V-bar transfer manoeuvre with in two orbital periods with FH-MPC

half the thrust such that it takes two orbits to reach the final state. The resulting ΔV is also the same as that obtained with (3.52), where Δx is half the total transfer distance.

A one-orbit R-bar transfer manoeuvre is presented in Fig. 4.17. It is observed that the resulting trajectory resembles the Hohmann transfer, such as the one presented in Fig. 3.9, and thus the manoeuvre is actually completed in only half of one

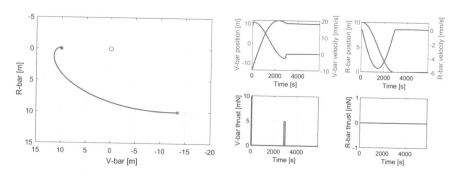

Fig. 4.17 R-bar transfer manoeuvre in half an orbital period with FH-MPC

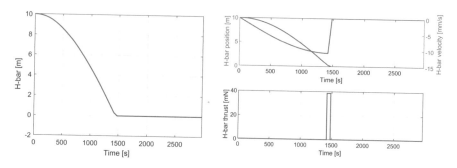

Fig. 4.18 H-bar transfer manoeuvre in half an orbital period with FH-MPC

orbit. The ΔV is also the same as that obtained with (3.50), although note that the FH-MPC manoeuvre is performed with asymmetric thrusts, generating a slightly different trajectory but equivalent regarding fuel consumption which suggests that the optimization problem is not strictly convex.

Next, in Fig. 4.18 an H-bar correction manoeuvre is performed in one orbital period. Because the ascending node is crossed after only half an orbit, the reference is reached after this period of time. The ΔV obtained is the same as that obtained for the inclination correction manoeuvre in Fig. 3.12 and computed with (3.53).

Introducing now some eccentricity, the manoeuvre presented in Fig. 4.19 is in the same conditions as that in Fig. 3.21. The trajectory obtained with the FH-MPC formulation is the same as the ideal one with the two impulses, and the ΔV required is the same as that computed with (3.54) and (3.56), since the generated manoeuvre also only has two thrust actions. Note also that, because the target orbit is now elliptical, the eccentric anomaly sampling technique presented in Sect. 4.1 is used, where E_s is the sampling eccentric anomaly, meaning that the initial and final thrust actions have different durations.

To demonstrate the behaviour in a highly elliptical orbit, a manoeuvre in the conditions of the PROBA-3 RVX scenario is considered, with the target orbital parameters presented in Table 4.4. The manoeuvre is presented in Fig. 4.20, for which it can be

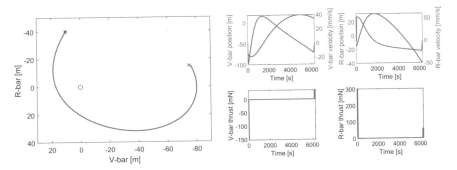

Fig. 4.19 Arbitrary in-plane transfer manoeuvre in half an orbital period in elliptic orbit with FH-MPC

Table 4.4 Target orbital parameters for PROBA-3 RVX scenario

Parameter	Value
Semi-major axis	36,941 km
Eccentricity	0.8111
Inclination	59°
RAAN	84°
Argument of periapsis	188°
Initial true anomaly	179°

seen from Table 4.3 that the resulting ΔV is 52.1 mm/s. On the other hand, the ΔV obtained for the ideal two impulse manoeuvre, which can be computed with (3.54) and (3.56), is 72.7 mm/s, which is significantly higher. The more efficient manoeuvre obtained with the FH-MPC formulation is due to the use of an intermediate thrust in V-bar at around 30,000 s, while the ideal manoeuvre is constrained to only initial and final thrusts. This greater degree of freedom from the use of MPC thus allows it to generate more fuel-efficient trajectories than the traditional two-impulse manoeuvres.

Finally, to demonstrate the inclusion of the H-bar motion, an actual manoeuvre to be performed for the PROBA-3 RVX is presented in Fig. 4.21. It can be observed from Table 4.3 that the required ΔV is 407.4 mm/s. On the other hand, the ΔV required for the two-impulse manoeuvre computed with (3.54) and (3.56) is 481 mm/s. Therefore, the FH-MPC formulation requires only 85% of the fuel required by a typical two-boost manoeuvre, which, once again, is achieved via intermediate thrust action, this time seen in H-bar. We have thus showed in this section through simulation that the FH-MPC formulation is fuel-optimal for a given manoeuvre duration.

Regarding computational load, notice from Table 4.3 that the increase of the prediction horizon has a relatively small impact on the execution time. For example, for a horizon of $N = 50$ the worst-case computation took 8.84 ms while increasing the horizon by four times to $N = 200$ resulted in this value being 12.1 ms, which

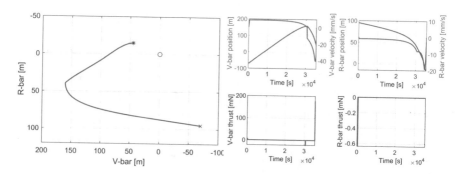

Fig. 4.20 Arbitrary in-plane transfer manoeuvre in half an orbital period in PROBA-3 orbit with FH-MPC

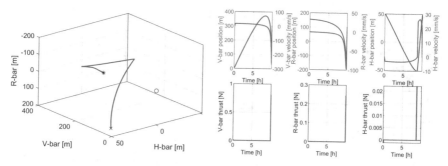

Fig. 4.21 Manoeuvre from the PROBA-3 RVX with FH-MPC

is only an approximately 37% increase. This is due to the fact that the optimization problem is formulated as a linear program, which can be solved very efficiently, in this case with the dual simplex algorithm. Furthermore, in these simulations, there aren't any additional control or state constraints, which would increase the computational load. In the case of control constraints, however, this increase would not be very significant.

Figure 4.22 shows the effect of the prediction horizon on the execution time, in the conditions of the PROBA-3 manoeuvre presented in Fig. 4.21, but with control saturation constraints of 1/3 N in each component and maintaining the manoeuvre duration constant. For each prediction horizon, the simulation was run ten times, and the execution times were averaged. It can be seen execution time grows approximately linearly, although the worst-case time oscillates unpredictably. This shows the computational advantage of formulating the optimization problem as an LP, where increasing the prediction horizon by a factor of 50 only increases the computation time by less than two and half times.

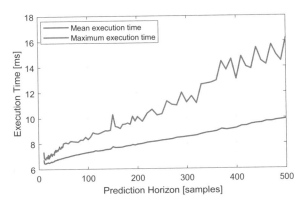

Fig. 4.22 Computation time of the PROBA-3 RVX manoeuvre as a function of the prediction horizon

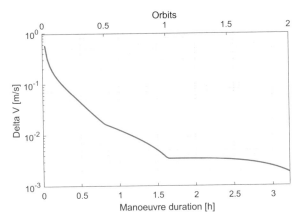

Fig. 4.23 V-bar transfer manoeuvre in circular orbit ΔV as a function of its duration

4.7.2 Variable Horizon Model Predictive Control

The VH formulation is useful to optimize both the manoeuvre duration and required fuel, where the trade-off between the two can be tuned with the parameter γ. If γ is zero, the solution will be the manoeuvre duration that minimizes the fuel only, where the duration is bounded by the maximum prediction horizon N_{max}. In simpler manoeuvres, however, the required fuel may be strictly decreasing with the manoeuvre duration.

Figure 4.23 shows the ΔV required for a V-bar transfer manoeuvre in a circular orbit as a function of the duration of the transfer. For a half-orbit transfer, the result corresponds to that in Fig. 4.15, with one orbit we get the result from Fig. 4.8, and a two-orbit transfer corresponds to the result from Fig. 4.16. It can be then seen that the required ΔV strictly decreases with the increase of the manoeuvre duration.

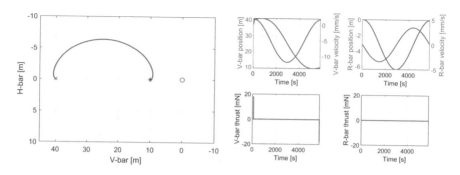

Fig. 4.24 V-bar transfer manoeuvre with VH-MPC, with a maximum transfer of one orbit and no manoeuvre duration cost

Table 4.5 Controller parameters and results for VH-MPC experiments

Figures	T_s	E_s	e	N_{max}	γ	N	ΔV (mm/s)	t_{max} (ms)	t_{avg} (ms)
4.24	98.1 s	–	0	60	0	60	3.45	138	54.9
4.25	98.1 s	–	0	60	0.01	30	16.90	184	121
4.27	57.9 s	–	0	100	0	32	130.9	412	277
4.29	–	3.6°	0.8111	100	0	41	48.16	710	595

Applying the VH-MPC formulation in the conditions from Fig. 4.23 with a maximum manoeuvre duration of one orbit and $\gamma = 0$ yields, the result from Fig. 4.24, where indeed the maximum transfer time is optimum. The parameters and results for these experiments are presented in Table 4.5.

If a manoeuvre duration cost is included ($\gamma > 0$), the optimal solution will have a shorter transfer time. In Fig. 4.25, the optimal transfer time becomes approximately half an orbital period, and the trajectory is similar to the ideal V-bar transfer with radial impulses presented in Fig. 3.10, although there is some actuation in V-bar. The total ΔV for that manoeuvre, determined with Eq. (3.52), is 16.3 mm/s, which is slightly lower than that obtained in this experiment due to the small V-bar actuation present there.

When the manoeuvre initial and reference points are not invariant, meaning that natural drift occurs if no thrust action is applied, then the manoeuvre ΔV may no longer be strictly decreasing with the duration. Figure 4.26 represents the ΔV as a function of the duration for a manoeuvre in such conditions, where the initial and final states are not on V-bar, and for a circular target orbit. It can be seen that the minimum transfer time within the two-orbit interval plotted is just over half an hour, at about 32% of an orbit. This happens because the chaser is drifting relative to the target, and thus there is an optimal time for performing the manoeuvre that is not the maximum time. When VH-MPC is applied in this scenario with a maximum duration

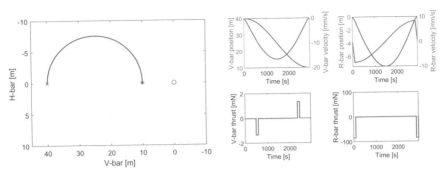

Fig. 4.25 V-bar transfer manoeuvre with VH-MPC, with a maximum transfer of one orbit and a manoeuvre duration cost

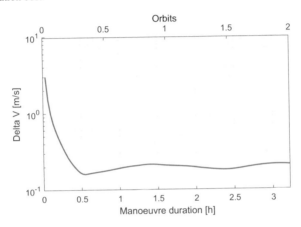

Fig. 4.26 Arbitrary in-plane transfer manoeuvre in circular orbit ΔV as a function of its duration

of one orbit, it yields a manoeuvre with a duration that is approximately 32% of an orbit, as shown in Fig. 4.27, thus validating the VH-MPC formulation.

We will now analyse the PROBA-3 manoeuvre from Fig. 4.21. Figure 4.28 shows the ΔV required for that manoeuvre as a function of the transfer time. It can again be seen that the ΔV is not strictly decreasing with the manoeuvre duration and that the peaks of this plot are much more pronounced than for the previous one. This happens because now the orbit is elliptical and the dynamics time-varying, meaning that there is another factor regarding the optimal time for performing the manoeuvre. For the PROBA-3 manoeuvre, there is a local minimum at around 40% of an orbit, but the globally optimum transfer time is one orbit, while there are peaks in the ΔV for half-orbit and one-and-a-half orbit transfers, the first of which corresponds to the manoeuvre duration from Fig. 4.21.

Applying VH-MPC to this manoeuvre with a maximum transfer time of one orbit yields the result from Fig. 4.29, where the optimal duration is 41% of an orbit. It can be seen from Table 4.5 that the ΔV required is about eight times less than that

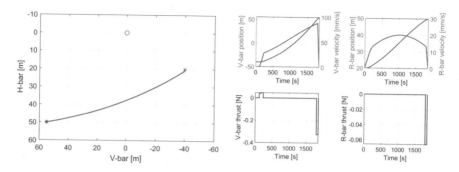

Fig. 4.27 Arbitrary in-plane transfer manoeuvre with VH-MPC, with a maximum transfer of one orbit and no manoeuvre duration cost

Fig. 4.28 PROBA-3 manoeuvre ΔV as a function of its duration

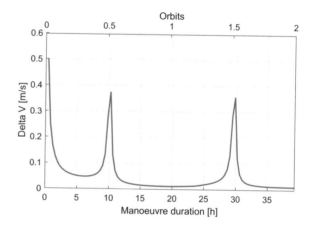

obtained in Fig. 4.21. Although this value does correspond to a minimum from the plot in Fig. 4.28, it is not the global minimum, which would be a duration of one orbital period. This is due to the fact that the algorithm used (branch and bound) does not optimize globally, and thus it converges to this local minimum. For online VH-MPC, this can be a problem as the optimization may converge to a different minimum than previous iterations, which could affect performance; this can be avoided with the use of warm start. For offline VH-MPC, it is feasible to optimize the problem globally.

We see from Table 4.5 that the computational load for the VH-MPC formulation is significantly higher than for FH-MPC, due to the fact that the first is a MILP and the latter an LP. It can then be infeasible to use this formulation online, especially with the inclusion of the very computationally heavy passive safety constraints.

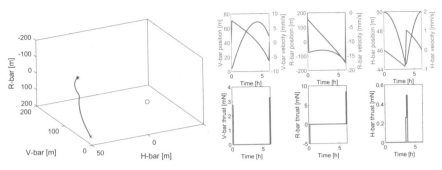

Fig. 4.29 PROBA-3 manoeuvre with VH-MPC, with a maximum transfer of one orbit and no manoeuvre duration cost

Table 4.6 Controller parameters and results for passive safety experiments

Figures	Parameters						Results					
	T_S	E_S	e	θ_0	N	S	ΔV	t_{max}	n_I	ΔV	t_{max}	n_I
4.30	193.4 s	–	0	–	30	30	1.616 mm/s	1.69 s	7	1.616 mm/s	81.1 ms	5
4.31	193.4 s	–	0	–	30	60	2.796 mm/s	6.97 s	8	INFEASIBLE		
4.32	–	7°	0.8111	245°	40	80	1.647 mm/s	33.7 s	4	1.646 mm/s	279 ms	5
4.33	–	3.89°	0.8111	30°	45	90	120.4 mm/s	132 s	3	119.9 mm/s	260 ms	3
							SQP			OASLP		

4.7.3 Passive Safety

This section features simulations of rendezvous manoeuvres with the passive safety constraint. The OASLP technique for obstacle avoidance presented in Sect. 4.5.3, which relies on a sequence of LPs, is utilized and compared with a standard SQP nonlinear optimization algorithm, similar to that done in [35]. An absolute ΔV iterative change of 10^{-5} m/s is used as the stopping criterion for the OASLP algorithm. The choice of this criterion is convenient since it allows for defining the required optimality in a quantitative way with regard to the ΔV The simulation parameters and results are presented in Table 4.6, where n_I denotes the number of OASLP iterations including the initial optimization with no obstacle constraints, or the number of SQP iterations. Similar to the OASLP algorithm, the first SQP MPC optimization is warm-started with the solution of the unconstrained problem for better convergence.

Figure 4.30 shows the result of a one-orbit V-bar transfer, similar to that in Fig. 4.10, with the inclusion of the passive safety constraint with a horizon of one orbital period. Although the failure trajectories from all discretized points in the trajectory are constrained, only the final is represented (in red), so as to not overbear the figure. The target spacecraft safety region is a 2-metre radius circle, that appears as an ellipse due to different scales being used. The thinner lines represent the results of intermediate OASLP iterations. The trajectories of the SQP and OASLP algorithms are superimposed, due to converging to the same solution, and thus are indiscernible in the figure.

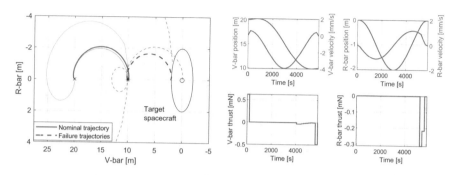

Fig. 4.30 One-orbit V-bar transfer manoeuvre in a circular orbit and with a one-orbit passive safety horizon. Lighter lines indicate the results of intermediate iterations

As shown in Table 4.6, the two algorithms converge to the same solution within a small relative ΔV difference (OASLP slightly lower), although the OASLP algorithm requires fewer iterations and is over 20 times faster. Regarding the trajectory, it is initially similar to that without the constraints, although toward the end of the manoeuvre, there is some extra non-sparse actuation, which results in a slight widening of the nominal trajectory such that the failure trajectories now stop, in the worst case, exactly on the edge of the safety region. The inclusion of the passive safety constraint results in a ΔV increase ΔV of approximately 17%, as is expected, and an increase of the computation time by nearly 10 times for OASLP and approximately 200 times for SQP.

In Fig. 4.31, the previous manoeuvre is repeated with an increased safety horizon of two orbital periods. For these conditions, the linear constraints determined for the OASLP algorithm yield an empty set, and thus the optimization becomes infeasible. This limitation will be tackled in future work, as mentioned previously in Sect. 4.5.4. Thus, the figure presents the result of the SQP algorithm instead.

The extra R-bar actuation before the 4000 s widens the approach in a such way that now the failure trajectories do not collide within the specified two orbits. The ΔV for this manoeuvre is almost double that of the unsafe one, and represents an increase of over 70% in respect to the one with a one orbit safety horizon. Furthermore, the SQP computation time has increased by over four times due to the additional constraints. It is remarked that as the safety horizon increases, the R-bar actuation does too. This results in the trajectory increasingly resembling the V-bar transfer manoeuvre with radial impulses, which, as mentioned previously in Sect. 4.5, guarantees passive safety in an infinite horizon, although it is over four times more costly.

Figure 4.32 presents an arbitrary in-plane transfer in the conditions of the PROBA-3 mission and with a safety horizon of approximately one orbit and a half, and with an increased safety region radius of 5 m. It can be seen that for the unconstrained trajectory the nominal trajectory also violates the safety region, besides the failure trajectory. The results of the OASLP and SQP algorithms are now more noticeably

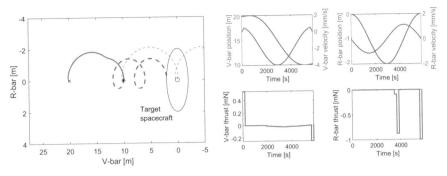

Fig. 4.31 V-bar transfer manoeuvre in one orbit, in a circular orbit and with a two-orbit passive safety horizon

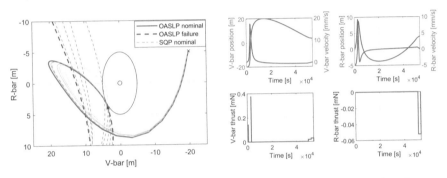

Fig. 4.32 Arbitrary in-plane transfer manoeuvre in PROBA-3 orbit and with approximately a one-and-a-half-orbit passive safety horizon

different, with the former being slightly more optimal and having a computation time two orders of magnitude faster.

Figure 4.33 presents a manoeuvre in all three dimensions, and thus the safety region is now a sphere, with a radius of 10 m. Because now the ΔV for the manoeuvre is two orders of magnitude greater than before, a tolerance of 10^{-3} m/s is used for the OASLP stopping criterion instead. As shown in the figure, the algorithms converge to slightly different local minima. Furthermore, as shown in Table 4.6, with only three iterations the OASLP algorithm satisfies its stopping criterion, resulting in a slightly more optimal manoeuvre than SQP, and with a computation time more than 500 times faster.

4.7.4 Robustness Experiments

To experiment and demonstrate the performance of the FH-MPC strategy in the presence of perturbations and disturbances, the algorithm has been implemented in a high-fidelity simulation environment developed by Deimos in the context of the

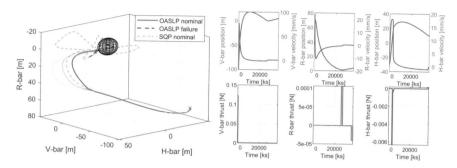

Fig. 4.33 Arbitrary in-plane transfer manoeuvre in PROBA-3 orbit and with approximately a one-and-a-half-orbit passive safety horizon

European Space Agency (ESA) CLGADR project.[1] Namely, for the following simulations, the controller will be subject to modelling errors, external perturbations, navigation errors and thruster errors. The robustness techniques covered in Sect. 4.6 will then be applied in order to improve the robustness of the guidance, regarding both feasibility and performance. Although there is typically also the need for guaranteeing robust constraint satisfaction for state constraints, such as the passive safety constraint, no techniques for that purpose are demonstrated here.

As previously mentioned, in this case, the MPC algorithm functions as a guidance algorithm, despite the fact that it operates in closed loop, since it runs at a very low frequency. Moreover, to test the algorithm performance independently from whichever high-frequency control strategy may be selected, no controller is added to the loop. Therefore, the results presented are not representative of real flight performance and merely serve to evaluate the MPC strategy and compare it to other guidance strategies available. For that same reason, the simulation does not include fully realistic models and conditions, such as thruster and sensor models, and all possible external perturbations, since many of these are typically compensated with the higher frequency controller.

To start with, only deterministic perturbations are included in the simulation. Namely, it includes the second-order geopotential model (J_2), third-body gravity from the Moon and Sun, and solar radiation pressure. Also note that even the first-order term of the geopotential model also constitutes a source of error for the guidance, since it uses the linearised Yamanaka–Ankersen. Control limits of 1 N are added in each thrust direction for this and all subsequent simulations. The MPC parameters and results for the following simulations are presented in Table 4.7, where e_{pos} and e_{vel} are the terminal errors in position and velocity.

A valid strategy for robustness to disturbances for MPC is to simply rely on its inherent robustness, due to it being a closed-loop strategy. Figure 4.34 exemplifies the

---

[1]This simulation environment was developed under ESA contract No. 4000111160/14/NL/MH. The views expressed herein can in no way be taken to reflect the official opinion of the European Space Agency.

Table 4.7 Controller parameters and results for robustness experiments with PROBA-3 manoeuvre

Figures	E_s	N	ΔV (mm/s)	e_{pos}	e_{vel}	t_{max} (ms)	t_{avg}
4.34	1.70°	100	407.2	36.92 m	2.344 cm/s	2.01	–
4.35	1.70°	100	132.6	9.580 m	269.7 mm/s	3.00	1.34 ms
4.36	1.70°	100	408.3	2.912 cm	0.416 mm/s	2.53	1.32 ms
4.37	0.85°	200	407.0	2.687 cm	0.499 mm/s	8.00	2.27 ms

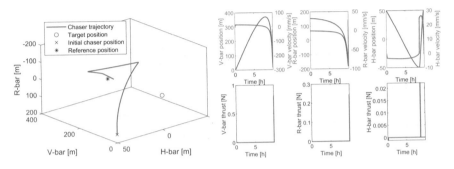

Fig. 4.34 PROBA-3 manoeuvre in high-fidelity simulator in open loop

performance of open-loop MPC in face of the mentioned perturbations, with the same PROBA-3 manoeuvre presented in Fig. 4.21. The chaser follows a similar trajectory as that in perfect conditions, having a similar total ΔV, but the perturbations not modelled in the guidance result in a very significant terminal error of 36.92 m for position and 2.344 cm/s for velocity.

Performing the same manoeuvre but in closed-loop yields, the result in Fig. 4.35. It can now be seen that the controller can now better approach the reference state despite the prediction model not being perfect, although the trajectory is slightly different than what was obtained in the unperturbed case. However, the optimization problem becomes infeasible in the final iteration due to the terminal constraint, and thus the manoeuvre ends with a position error of 9.58 m and a relative velocity error of 269.7 mm/s. Thus, as previously mentioned in Sect. 4.6.2, this formulation suffers from robust feasibility issues. Although a solution to an infeasible iteration may be to use the second time-step of the previous iteration, it is preferable to avoid infeasibility altogether.

4.7.4.1 Robust Feasibility

To deal with infeasibility in the terminal state constraint, the feasible terminal box technique presented in Sect. 4.6.2 is applied. This yields the result from Fig. 4.36, where the optimization problem again becomes feasible at the final iteration, allow-

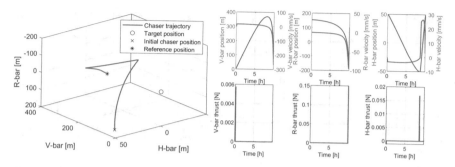

Fig. 4.35 PROBA-3 manoeuvre in high-fidelity simulator in closed loop

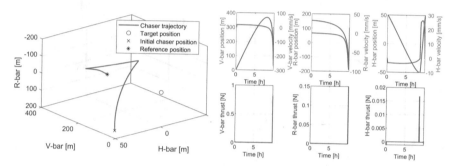

Fig. 4.36 PROBA-3 manoeuvre in high-fidelity simulator with closed-loop and feasible terminal box

ing for the manoeuvre to be completed, without any significant computational load increase. Furthermore, as shown in Table 4.7, the residual errors in position and velocity are of 2.912 cm and 0.416 mm/s, respectively, despite all the perturbations present and not modelled in the guidance. These errors are very small when compared to the result in open loop and given all the perturbations present, therefore demonstrating the inherent MPC performance robustness. Furthermore, this improved performance by running in closed loop comes at no significant additional ΔV cost with respect to the ideal value, despite all the existing perturbations.

In Fig. 4.37, the prediction horizon is doubled, while maintaining the manoeuvre duration. As shown in Fig. 4.34, this increase has no significant improvement on the terminal state error or the ΔV. On the other hand, the computational time has also doubled on average, and nearly quadrupled in the worst case.

4.7.4.2 Robust Performance

The fuel performance and accuracy of the FH-MPC strategy will now be analysed in the presence of other disturbances, namely navigation and actuator errors, and improved with the techniques proposed in Sect. 4.6. The following navigation performance model is used:

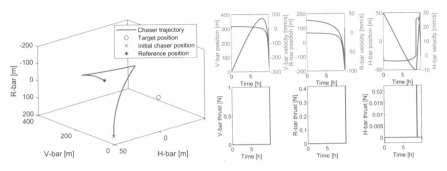

Fig. 4.37 PROBA-3 manoeuvre in high-fidelity simulator with feasible terminal box and increased prediction horizon

Table 4.8 Controller parameters and results for stochastic robustness experiments with PROBA-3 manoeuvre

Figures	E_S	N	$\varepsilon_{0,p}$	$\varepsilon_{0,v}$	N_T	N_r	ΔV[mm/s]		e_{pos}[cm]		e_{vel}[mm/s]		t_{max}[ms]	
							μ	σ	μ	σ	μ	σ	μ	σ
4.38	1.70°	100	–	–	–	1	524	20.0	35.5	15.1	3.91	1.77	4.15	0.813
4.39	1.68°	200	–	–	–	1	713	25.2	27.4	13.4	3.92	1.72	7.05	0.887
4.40	1.70°	100	15 m	10 mm/s	–	1	425	15.5	2310	109	3.53	1.49	6.74	0.872
4.41	1.70°	100	15 m	10 mm/s	–	1	479	15.6	55.4	10.2	3.87	1.73	6.65	0.913
4.42	1.70°	100	15 m	10 mm/s	10	1	485	18.0	20.0	9.03	3.35	1.45	5.27	0.831
4.43	1.70°	100	15 m	10 mm/s	10	5	447	15.7	20.9	9.05	3.42	1.43	5.39	0.723

- Relative position accuracy (3-σ): 30 cm,
- Relative velocity accuracy (3-σ): 3 mm/s.

Although these values may seem inconsequential, such a degree of uncertainty in the estimated state may imply a dispersion of hundreds of meters after half an orbit. Thruster errors are also included in the simulation with the following performance model which, besides modelling magnitude and direction errors, includes a thrust threshold for which commands are not executed in order to model the effect of the thrusters' minimum impulse bit (MIB):

- Thrust magnitude accuracy (3-σ): 30%,
- Thrust direction accuracy (3-σ): 3°,
- Cut-off thrust due to MIB: 1 mN.

Since the simulation now has stochastic perturbations, the following tests are performed with Monte Carlo campaigns of 20 repetitions. The mean (μ) and standard deviation (σ) are computed for the results and presented in Table 4.8.

In the same conditions as in Fig. 4.37 but with the above navigation and actuation errors yields the results in Fig. 4.38. The dispersion between the trajectories can be observed, although for all of them the guidance is able to converge on the reference. The right plots correspond to only one of the simulations. It can be observed that the

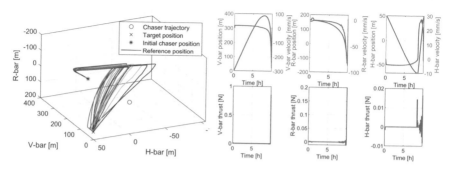

Fig. 4.38 PROBA-3 manoeuvre in high-fidelity simulator, with navigation and actuation errors

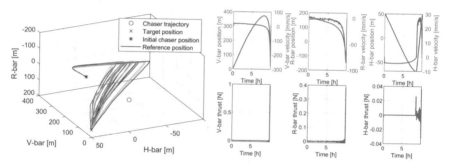

Fig. 4.39 PROBA-3 manoeuvre in high-fidelity simulator, navigation and actuation errors, and increased prediction horizon

control action is now less sparse since now the controller corrects the trajectory at every step in an attempt to satisfy the terminal state constraint, since every iteration is performed with imperfect knowledge of the state, and each control action is applied with errors. This results in a mean increase in ΔV of 41% with respect to without navigation and actuation errors, which is very significant. Furthermore, the mean final state errors have increased by one order of magnitude.

In an attempt to improve the performance, the number of samples in the prediction horizon is doubled, yielding the result Fig. 4.39. The trajectories are now slightly less dispersed and the residual error is smaller, although the average ΔV has increased to almost double that without navigation and actuation errors. This is due to the fact that with an increased number of MPC samples the guidance performs more correction manoeuvres, which are planned based on imperfect information, and thus more fuel is wasted in attempts to drive the predicted terminal state exactly to the reference. Furthermore, these corrections are not performed as planned, due to actuator errors, inciting further corrections. Thus, increasing the prediction horizon in the presence of stochastic navigation and actuator errors increases the required fuel, which is opposite to what may be typically expected with non-stochastic perturbations.

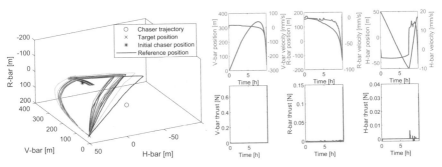

Fig. 4.40 PROBA-3 manoeuvre in high-fidelity simulator, navigation and actuation errors and fixed terminal box

4.7.4.3 Dynamic Terminal Box

To decrease the controller sensitivity to stochastic disturbances, we substitute the terminal state constraint with a terminal box, as discussed in Sect. 4.6.3. Using an 15 m terminal box for position ($\varepsilon_{0,p}$) and 10 mm/s for velocity ($\varepsilon_{0,v}$), and maintaining these dimensions constant throughout the manoeuvre, yields the result from Fig. 4.40. There are now fewer manoeuvre corrections, which results in a significant decrease of 99 mm/s in respect to the result from Fig. 4.38. However, the mean terminal error for position is significantly higher at 23.1 m, which is due to the loosening of the terminal constraint.

To maintain accuracy and still reduce sensitivity to disturbances, the dynamic terminal box approach presented in Sect. 4.6.3 is now used, with a linear decrease with time of the terminal box dimensions. Using the previous box dimension as the initial values yields the result from Fig. 4.41. The ΔV has increased from the previous simulation, although it remains below that without the terminal box in Fig. 4.38, at no cost of extra computational complexity. Furthermore, the terminal error for position and velocity has improved in comparison to the result from Fig. 4.38. Although the gain in ΔV is not significant, this result validates this type of approach for achieving better fuel performance in the presence of stochastic disturbances. With further parameter tuning and a different method for varying the terminal box dimensions, even better performance might be achieved.

4.7.4.4 Quadratic Terminal Controller

Finally, we note that the terminal state error at the end of the manoeuvre is still significant, namely for position, which has a mean value of 55.4 cm for the previous simulation. This value, namely position, is still significantly higher than the navigation uncertainty and thus can be improved. As mentioned in Sect. 4.6.4, the terminal errors are due to the sparsity and fixed final-time of the FH-MPC controller. Thus, we attempt to improve this by employing the use of a terminal quadratic controller that

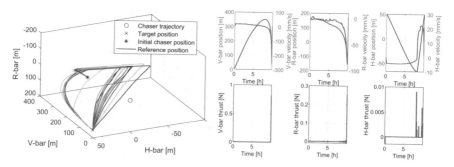

Fig. 4.41 PROBA-3 manoeuvre in high-fidelity simulator, navigation and actuation errors and dynamic terminal box

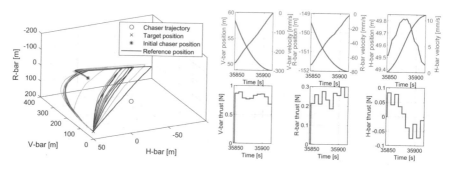

Fig. 4.42 PROBA-3 manoeuvre in high-fidelity simulator, navigation and actuation errors, dynamic terminal box and quadratic terminal controller

substitutes the final iteration of the nominal controller, as described in that section. With a prediction horizon of $N_T = 10$, an input cost matrix of $R = I$ and terminal state cost matrix $Q_f = \text{diag}[1, 1, 1, 10^4, 10^4, 10^4]$ yields the result from Fig. 4.42, where the thrust profile of the quadratic controller is shown.

A small increase in the ΔV can be seen, while the mean residual position error decreases to 20 cm. The terminal controller only operates during what would instead be the last sample of the FH-MPC controller, and thus incurs no increase in manoeuvre duration. Better performance might be achieved with a further tuning of the terminal horizon N_T or the terminal cost matrices R and Q_f, although this result validates this approach of achieving better accuracy.

4.7.4.5 Re-solve Rate

An additional parameter available which can decrease the sensitivity of the controller to disturbances is the re-solve rate of the MPC optimization. If the problem is not re-solved at every single time-step, the controller will then perform fewer over-corrections along with the manoeuvre, and potentially expend less fuel. On the other

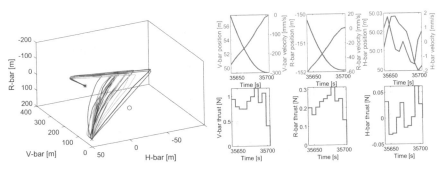

Fig. 4.43 PROBA-3 in high-fidelity simulator, navigation and actuation errors, dynamic terminal box and quadratic terminal controller, and lower re-solve rate

hand, the terminal position and velocity errors will be expectedly worse with fewer corrections. To minimize this last effect, the terminal quadratic controller is still re-solved at every MPC time-step.

For $N_r = 5$, meaning the MPC problem is only re-solved every five time-steps, the result in Fig. 4.43 is obtained. The total mean manoeuvre ΔV has decreased to 447 mm/s, which represents an increase of only 9.77% with respect to the ideal value in the absence of any perturbations and disturbances. Furthermore, since the quadratic controller is still re-solved at every MPC time-step, the terminal error remains approximately the same as in the previous simulation with $N_r = 1$.

References

1. W. Fehse, *Automated Rendezvous and Docking of Spacecraft* (Cambridge University Press, 2003). ISBN: 0521824923
2. E.N. Hartley, A tutorial on model predictive control for spacecraft rendezvous, in *2015 European Control Conference (ECC)*, July 2015 (2015), pp. 1355–1361
3. A. Richards, J. How, Performance evaluation of rendezvous using model predictive control, in *AIAA Guidance, Navigation, and Control Conference and Exhibit*, November 2003 (2003)
4. S. Di Cairano, H. Park, I. Kolmanovsky, Model predictive control approach for guidance of spacecraft rendezvous and proximity maneuvering. Int. J. Robust Nonlinear Control **22**, 1398–1427 (2012)
5. E.N. Hartley, P.A. Trodden, A.G. Richards, J.M. Maciejowski, Model predictive control system design and implementation for spacecraft rendezvous. Control Eng. Pract. **20**, 695–713 (2012)
6. L. Breger, J. How, A. Richards, Model predictive control of spacecraft formations with sensing noise, in *Proceedings of the 2005, American Control Conference* (2005), pp. 2385–2390
7. F. Gavilan, R. Vazquez, E.F. Camacho, Chance-constrained model predictive control for spacecraft rendezvous with disturbance estimation. Control Eng. Pract. **20**, 111–122 (2012)
8. A. Weiss, M. Baldwin, R.S. Erwin, I. Kolmanovsky, Model predictive control for spacecraft rendezvous and docking: strategies for handling constraints and case studies. IEEE Trans. Control Syst. Technol. **23**, 1638–1647 (2015)
9. R. Vazquez, F. Gavilan, E.F. Camacho, Model predictive control for spacecraft rendezvous in elliptical orbits with on/off thrusters. IFAC-PapersOnLine **48**, 251–256 (2015)

10. S. Zhu, R. Sun, J. Wang, J. Wang, X. Shao, Robust model predictive control for multi-step short range spacecraft rendezvous. Adv. Space Res. **62**, 111–126 (2018)
11. Ø. Hegrenæs, J. Gravdahl, P. Tøndel, Spacecraft attitude control using explicit model predictive control. Automatica **41**, 2107–2114 (2005). (Dec.)
12. V. Manikonda, P.O. Arambel, M. Gopinathan, R.K. Mehra, F.Y. Hadaegh, A model predictive control-based approach for spacecraft formation keeping and attitude control, in *Proceedings of the 1999 American Control Conference (Cat. No. 99CH36251)*, vol. 6, June 1999 (1999), pp. 4258–4262
13. J. Lilja, Combined attitude and orbital MPC for thruster based spacecrafts (2017)
14. A. Walsh, S. Di Cairano, A. Weiss, MPC for coupled station keeping, attitude control, and momentum management of low-thrust geostationary satellites, July 2016 (2016), pp. 7408–7413
15. J.C. Sanchez, F. Gavilan, R. Vazquez, C. Louembet, Aflatness-based predictive controller for six-degrees of freedom spacecraft rendezvous. Acta Astronaut. **167**, 391–403 (2020)
16. I.O. Burak, Model predictive control for spacecraft rendezvous and docking with uncooperative targets. Ph.D. thesis, Nanyang Technological University (2020)
17. P. Rosa et al., Autonomous close-proximity operations in space: the PROBA-3 rendezvous experiment (P3RVX), in *69th International Astronautical Congress (IAC 2018)* (2018)
18. K. Yamanaka, F. Ankersen, New state transition matrix for relative motion on an arbitrary elliptical orbit. J. Guid. Control Dyn. **25**, 60–66 (2002)
19. F. Ankersen, Guidance, navigation, control and relative dynamics for spacecraft proximity maneuvers. Ph.D. thesis, Institut for Elektroniske Systemer (2010). ISBN: 9788792328724
20. K. Alfriend, H. Yan, Evaluation and comparison of relative motion theories. J. Guid. Control Dyn. **28**, 254–261 (2005)
21. J. Sullivan, S. Grimberg, S. D'Amico, Comprehensive survey and assessment of spacecraft relative motion dynamics models. J. Guid. Control Dyn. **40**, 1837–1859 (2017)
22. C. Wei, S.-Y. Park, C. Park, Linearized dynamics model for relative motion under a J2-perturbed elliptical reference orbit. Int. J. Non-Linear Mech. **55**, 55–69 (2013)
23. L. Cao, A.K. Misra, Linearized J2 and atmospheric drag model for satellite relative motion with small eccentricity. Proc. Inst. Mech. Eng. Part G: J. Aerosp. Eng. **229**, 2718–2736 (2015)
24. H. Schaub, S.R. Vadali, J.L. Junkins, K.T. Alfriend, Spacecraft formation flying control using mean orbit elements. J. Astronaut. Sci. **48**, 69–87 (2000)
25. K. Alfriend, Nonlinear considerations in satellite formation flying, in *AIAA/AAS Astrodynamics Specialist Conference and Exhibit* (2002), p. 4741
26. D.-W. Gim, K.T. Alfriend, Satellite relative motion using differential equinoctial elements. Celest. Mech. Dyn. Astron. **92**, 295–336 (2005)
27. L. Breger, J.P. How, Gauss's variational equation-based dynamics and control for formation flying spacecraft. J. Guid. Control Dyn. **30**, 437–448 (2007)
28. S. D'Amico, Relative orbital elements as integration constants of Hill's equations. DLR, TN, 05-08 (2005)
29. O. Montenbruck, M. Kirschner, S. D'Amico, S. Bettadpur, E/I-vector separation for safe switching of the GRACE formation. Aerosp. Sci. Technol. **10**, 628–635 (2006)
30. S. D'Amico, Autonomous formation flying in low earth orbit. Ph.D. thesis, TU Delft (2010)
31. A.W. Koenig, T. Guffanti, S. D'Amico, New state transition matrices for spacecraft relative motion in perturbed orbits. J. Guid. Control Dyn. **40**, 1749–1768 (2017)
32. J.C. Sanchez, F. Gavilan, R. Vazquez, Chance-constrained model predictive control for near rectilinear halo orbit spacecraft rendezvous. Aerosp. Sci. Technol. **100**, 105827 (2020)
33. P. Lu, X. Liu, Autonomous trajectory planning for rendezvous and proximity operations by conic optimization. J. Guid. Control Dyn. **36**, 375–389 (2013)
34. A. Richards, J. How, Model predictive control of vehicle maneuvers with guaranteed completion time and robust feasibility, vol 5 (2003), pp. 4034–4040. ISBN: 0-7803-7896-2
35. C. Jewison, R.S. Erwin, A. Saenz-Otero, Model predictive control with ellipsoid obstacle constraints for spacecraft rendezvous. IFAC-PapersOnLine **48**, 257–262 (2015)

36. L. Ravikumar, R. Padhi, N. Philip, Trajectory optimization for rendezvous and docking using nonlinear model predictive control. IFAC-PapersOnLine **53**, 518–523 (2020)

37. A. Richards, E. Feron, J. How, T. Schouwenaars, Spacecraft trajectory planning with avoidance constraints using mixed-integer linear programming. J. Guid. Control Dyn. **25** (2002)

38. L.S. Breger, J.P. How, Safe trajectories for autonomous rendezvous of spacecraft. J. Guid. Control Dyn. **31**, 1478–1489 (2008)

39. J.B. Mueller, R. Larsson, Collision avoidance maneuver planning with robust optimization, in *7th International ESA Conference on Guidance, Navigation and Control Systems*, June 2008, Tralee, County Kerry, Ireland (2008)

40. Y. Mao, M. Szmuk, B. Açıkmeşe, Successive convexification of non-convex optimal control problems and its convergence properties, in *2016 IEEE 55th Conference on Decision and Control (CDC)* (2016), pp. 3636–3641

41. J. Nocedal, S. Wright, *Numerical Optimization* (Springer Science & Business Media, 2006)

42. F. Augugliaro, A.P. Schoellig, R. D'Andrea, Generation of collision-free trajectories for a quadrocopter fleet: a sequential convex programming approach, in *2012 IEEE/RSJ International Conference on Intelligent Robots and Systems* (2012), pp. 1917–1922

43. G. Deaconu, C. Louembet, A. Théron, Designing continuously constrained spacecraft relative trajectories for proximity operations. J. Guid. Control Dyn. **38**, 1208–1217 (2014)

44. J. Rawlings, D. Mayne, M. Diehl, *Model Predictive Control: Theory, Computation, and Design*, 2nd edn. (Nob Hill Publishing, 2017)

45. P.O. Scokaert, D. Mayne, Min-max feedback model predictive control for constrained linear systems. IEEE Trans. Autom. Control **43**, 1136–1142 (1998)

46. W. Langson, I. Chryssochoos, S. Raković, D.Q. Mayne, Robust model predictive control using tubes. Automatica **40**, 125–133 (2004)

47. J.P. How, M. Tillerson, Analysis of the impact of sensor noise on formation flying control, in *Proceedings of the 2001 American Control Conference (Cat. No. 01CH37148)*, vol 5 (2001), pp. 3986–3991

48. G. Deaconu, C. Louembet, A. Théron, Minimizing the effects of navigation uncertainties on the spacecraft rendezvous precision. J. Guid. Control Dyn. **37**, 695–700 (2014)

49. M. Mammarella, E. Capello, H. Park, G. Guglieri, M. Romano, Tube-based robust model predictive control for spacecraft proximity operations in the presence of persistent disturbance. Aerosp. Sci. Technol. **77**, 585–594 (2018)

50. K. Dong, J. Luo, Z. Dang, L. Wei, Tube-based robust output feedback model predictive control for autonomous rendezvous and docking with a tumbling target. Adv. Space Res. **65**, 1158–1181 (2020)

51. C. Louembet, D. Arzelier, G. Deaconu, Robust rendezvous planning under maneuver execution errors. J. Guid. Control Dyn. **38**, 76–93 (2014)

52. A.G. Richards, Robust constrained model predictive control. Ph.D. thesis, Massachusetts Institute of Technology (2005)

53. M. Tillerson, G. Inalhan, J.P. How, Co-ordination and control of distributed spacecraft systems using convex optimization techniques. Int. J. Robust Nonlinear Control: IFAC-Affil. J. **12**, 207–242 (2002)

Chapter 5
Conclusions and Future Work

This work has demonstrated MPC to be an appropriate method for implementing in a G&C system for orbital rendezvous. By using a cost function proportional to the fuel spent, a terminal state constraint, and the fixed-horizon (FH) strategy, a fuel-optimal formulation is obtained, given a pre-defined manoeuvre duration. Furthermore, this cost function can be re-written as a linear one, and the relative dynamics between the spacecraft can be accurately linearised via the Yamanaka–Ankersen state transition matrix [1], allowing for the use of a linear prediction model. This fuel-optimal formulation known as FH-MPC [2] becomes a linear program, which can be optimized very efficiently and thus possibly enabling real-time use. The formulation is not constrained to using this specific dynamic model, which can be easily swapped about with any linearized model, such as models which include disturbances [3, 4] or relative orbital element models [5, 6], by simply replacing the state model matrices.

The FH-MPC formulation presents advantages in fuel consumption over traditional guidance techniques, that usually rely on open-loop two-impulse manoeuvres to achieve a re-configuration between states. Since MPC is not constrained to just two control actions, intermediate thrust actions are naturally exploited by the controller to generate more efficient trajectories that are not possible with the two-impulse approach. This approach becomes especially advantageous for manoeuvres in highly elliptical orbits, such as the PROBA-3 mission rendezvous experiment (RVX) [7], where the dynamics are time-varying and the instants at which burns are executed become more critical. However, the FH-MPC formulation only optimizes the fuel for a given pre-defined transfer duration. The VH-MPC formulation allows thus optimizing both manoeuvre duration and fuel, by formulating the problem as a MILP [2]. Although the online use of MILP may be infeasible, VH-MPC can be used to determine offline the optimal manoeuvre duration, then executing the manoeuvre with FH-MPC.

© The Author(s), under exclusive license to Springer Nature Switzerland AG 2021
A. Botelho et al., *Predictive Control for Spacecraft Rendezvous*,
SpringerBriefs in Applied Sciences and Technology,
https://doi.org/10.1007/978-3-030-75696-3_5

One contribution from this book emerged from the fact that it addresses orbit eccentricities far higher than what is typically considered in the literature, but which are still relevant, as evidenced by the PROBA-3 mission. With the increase of the eccentricity, the dynamics also become increasingly more time-variant, such that these may be tens of times faster at perigee than at apogee. To tackle this, we have proposed to sample the MPC dynamics with constant eccentric anomaly intervals, rather than the widely used constant time intervals, such that the samples are spread more evenly along the orbit.

In this work, the Ankersen zero-order hold the particular solution [8] has been employed, while typically impulsive models are used for rendezvous with MPC, except for a few examples such as [9, 10]. A constant thrust parametrization models the spacecraft thrust profile more realistically, especially in close proximity manoeuvres, and eliminates a common source of error which lies in the discretization of the impulsive manoeuvres into a force profile subsequently to the manoeuvre being computed. Because the sampling intervals are very large, however, it may be undesirable to command constant thrust actions for such long periods of time. Nonetheless, this drawback may be overcome by defining, in the particular solution, the thrust duration to be less than the full MPC sampling period, in what would become a partial zero-order hold discretization. Moreover, most spacecraft orbital manoeuvring systems only have non-throttleable on/off thrusters, and thus perform intermediate thrust via pulse-width modulation (PWM). An even more realistic model is then to use the PWM parameters as decision variables, although this approach results in a nonlinear prediction model. In [9], the authors Vazquez et al. overcome this via a sequence of two linear programs.

The main advantage of MPC over other G&C strategies is that it allows for the explicit incorporation of control and state constraints [11]. In a rendezvous scenario, this possibility is useful to model thrust limitations, that can be accomplished realistically with linear constraints if the spacecraft has omnidirectional thrust, thus maintaining the FH-MPC formulation as a LP.

Another crucial operational constraint in close-proximity operations is related to passive collision safety [12]. This approach requires, for each failure trajectory, the addition of several obstacle avoidance constraints, that are naturally non-convex and thus yields a higher computational cost in the optimization. These constraints have previously been formulated as linear constraints [12–16], though with limited applicability. Thus, this book proposes a new approach for achieving collision avoidance with linear constraints, named Obstacle Avoidance with Sequential Linear Programming (OASLP). It relies on performing a sequence of LP optimizations with successively linearized and time-varying obstacle constraints, such that the LP solutions converge to the problem with the original nonlinear constraints. Simulation results showed this algorithm to be superior to a standard Sequential Quadratic Programming (SQP) algorithm, which showed computational times up to two orders of magnitude greater. Although simulations showed that the OASLP converges after a small number of iterations, it has not been theoretically proven that convergence actually occurs. Furthermore, there are scenarios where the LP sub-problem may become infeasible, a drawback that limits the use of this technique, and that must be tackled in future work. However, the satisfaction of non-convex constraints with convex optimization is desirable and promising, and therefore this technique warrants further research.

Another crucial aspect in the design of a G&C system is robustness with respect to disturbances and perturbations, of which the most significant in a rendezvous scenario are modelling errors, navigation uncertainty and execution errors. Although MPC can inherently achieve robust convergence [11], due to the fact that it is a closed-loop strategy, additional techniques might be required to achieve robust feasibility, performance and constraint satisfaction. To ensure that the terminal constraint present in the FH-MPC formulation never renders the optimization problem infeasible, we have utilized the feasible terminal box approach [17], that relaxes the terminal constraint to the minimum size that allows the problem to be feasible, while maintaining the problem as an LP and not significantly increasing the computational complexity.

Robust performance, which here refers to robust fuel performance and manoeuvre accuracy, is often not addressed in the literature. For avoiding over-correction of the trajectory due to stochastic disturbances, this book proposes a dynamically variable terminal box constraint that replaces the equality terminal state constraint, yielding better fuel performance and maintaining accuracy. This method, however, requires some tuning and should be the subject of further research. For example, the size of the terminal box may be computed as a function of the known uncertainties. The book also proposes the use of a quadratic terminal controller, that substitutes the final iteration of the FH-MPC controller, in order to allow for a more precise braking manoeuvre and granting better terminal accuracy. Neither of these proposed techniques increases the computational complexity of the problem. The book also presented the tuning of the MPC re-solve rate as a way for achieving better robust performance, namely regarding the fuel consumption.

These techniques have been validated in an industrial high-fidelity simulator, which includes gravitational perturbations, solar radiation pressure, navigation disturbances and actuation errors, including the thrusters' minimum impulse bit (MIB) effect, and considering the PROBA-3 RVX as the test scenario. The limitations of the basic FH-MPC in the presence of perturbation have been demonstrated, namely in regard to robust feasibility and performance. The robustness techniques presented were shown to improve these shortcomings, with no significant additional computational complexity. Namely, it has been demonstrated that these techniques allow the FH-MPC to robustly achieve the reference state with terminal errors comparable to the navigation disturbances, and with a total manoeuvre ΔV less than 10% higher than the ideal value, despite all the perturbations present and not modelled in the controller. Furthermore, it has been shown that the computational times obtained are not prohibitive of a real-time implementation.

Finally, although robust constraint satisfaction techniques were not considered here, the need for these techniques may be necessary for the passive safety constraint. The chance-constrained MPC approach presented in [18, 19] is a good candidate since it employs constraint tightening with online uncertainty estimation, and does not significantly increase computational complexity.

Although MPC has the potential to be used for implementing the G&C system for real rendezvous missions, its feasible implementation depends on the specific hardware for that mission. Furthermore, while MPC can offer increased autonomy and better fuel consumption profiles, this feature must be balanced with its greatest downside, which is the required computational complexity. Also, before it becomes an attractive alternative, there is a need to analyse whether MPC indeed offers fuel benefits over the traditional approach, taking into account all disturbances present in a rendezvous mission that can greatly reduce its performance. This ties into the need for a standard approach for ensuring robustness that remains feasible to implement in real time, which does not yet exist and thus calls for further research.

5.1 Open Research Topics

Finally, we summarize open research topics suggested along with this monograph:

- Consider and compare alternative prediction models, e.g. based on relative orbital elements and other dynamics, such as the three-body problem e.g. near-rectilinear halo orbits.
- Study convergence of the OASLP algorithm to a local minimum of the original nonlinear problem, address the cases where the linear constraints render the problem infeasible and provide boundedness proof.
- Test the passive safety constraint in the presence of disturbances and experiment with robust constraint satisfaction techniques.
- Study the inclusion of mid-thrust failures in the passive safety constraint, and ensuring the continuous-time trajectory does not violate the constraint.
- Improve the methods presented here for robust performance, namely the dynamic terminal box and the terminal quadratic controller.
- Test the MPC algorithms in a realistic simulation environment with 6-DoF dynamics and the low-level controller, to quantitatively evaluate performance and compare to traditional approaches.
- Implement and test the present MPC algorithms in an embedded environment comparable to flight hardware and software, utilizing optimization algorithms with guaranteed convergence and good early termination properties.

References

1. K. Yamanaka, F. Ankersen, New state transition matrix for relative motion on an arbitrary elliptical orbit. J. Guid. Control Dyn. **25**, 60–66 (2002)
2. A. Richards, J. How, Performance evaluation of rendezvous using model predictive control, in *AIAA Guidance, Navigation, and Control Conference and Exhibit*, November 2003 (2003)
3. C. Wei, S.-Y. Park, C. Park, Linearized dynamics model for relative motion under a J2-perturbed elliptical reference orbit. Int. J. Non-Linear Mech. **55**, 55–69 (2013)

4. L. Cao, A.K. Misra, Linearized J2 and atmospheric drag model for satellite relative motion with small eccentricity. Proc. Inst. Mech. Eng. Part G: J. Aerosp. Eng. **229**, 2718–2736 (2015)
5. L. Breger, J.P. How, Gauss's variational equation-based dynamics and control for formation flying spacecraft. J. Guid. Control Dyn. **30**, 437–448 (2007)
6. A.W. Koenig, T. Guffanti, S. D'Amico, New state transition matrices for spacecraft relative motion in perturbed orbits. J. Guid. Control Dyn. **40**, 1749–1768 (2017)
7. P. Rosa et al., Autonomous close-proximity operations in space: the PROBA-3 rendezvous experiment (P3RVX) in *69th International Astronautical Congress* (IAC 2018) (2018)
8. F. Ankersen, Guidance, navigation, control and relative dynamics for spacecraft proximity maneuvers. Ph.D. thesis, Institut for Elektroniske Systemer (2010). ISBN: 9788792328724
9. R. Vazquez, F. Gavilan, E.F. Camacho, Model predictive control for spacecraft rendezvous in elliptical orbits with on/off thrusters. IFAC-PapersOnLine **48**, 251–256 (2015)
10. R. Vazquez, F. Gavilan, E.F. Camacho, Pulse-width predictive control for LTV systems with application to spacecraft rendezvous. Control Eng. Pract. **60**, 199–210 (2017)
11. J. Rawlings, D. Mayne, M. Diehl, *Model Predictive Control: Theory, Computation, and Design*, 2nd edn. (Nob Hill Publishing, 2017)
12. L.S. Breger, J.P. How, Safe trajectories for autonomous rendezvous of spacecraft. J. Guid. Control Dyn. **31**, 1478–1489 (2008)
13. J.B. Mueller, R. Larsson, Collision avoidance maneuver planning with robust optimization in *7th International ESA Conference on Guidance, Navigation and Control Systems*, June 2008, Tralee, County Kerry, Ireland (2008)
14. S. Di Cairano, H. Park, I. Kolmanovsky, Model predictive control approach for guidance of spacecraft rendezvous and proximity maneuvering. Int. J. Robust Nonlinear Control **22**, 1398–1427 (2012)
15. E.N. Hartley, P.A. Trodden, A.G. Richards, J.M. Maciejowski, Model predictive control system design and implementation for spacecraft rendezvous. Control Eng. Pract. **20**, 695–713 (2012)
16. A. Weiss, M. Baldwin, R.S. Erwin, I. Kolmanovsky, Model predictive control for spacecraft rendezvous and docking: strategies for handling constraints and case studies. IEEE Trans. Control Syst. Technol. **23**, 1638–1647 (2015)
17. M. Tillerson, G. Inalhan, J.P. How, Co-ordination and control of distributed spacecraft systems using convex optimization techniques. Int. J. Robust Nonlinear Control: IFAC-Affil. J. **12**, 207–242 (2002)
18. F. Gavilan, R. Vazquez, E.F. Camacho, Chance-constrained model predictive control for spacecraft rendezvous with disturbance estimation. Control Eng. Pract. **20**, 111–122 (2012)
19. J.C. Sanchez, F. Gavilan, R. Vazquez, Chance-constrained model predictive control for near rectilinear halo orbit spacecraft rendezvous. Aerosp. Sci. Technol. **100**, 105827 (2020)

Printed in the United States
by Baker & Taylor Publisher Services